THE MELANIN BOOK

Electrician Manual

Chase Duquesnay
Enqi Real

Amazon

Copyright © 2024 Chase Duquesnay

All rights reserved

The characters and events portrayed in this book are fictitious. Any similarity to real persons, living or dead, is coincidental and not intended by the author. No parts of these books or book series are intended as Medical Advice simply edutainment

No part of this book may be reproduced, or stored in a retrieval system, or transmitted in any form or by any means, electronic, mechanical, photocopying, recording, or otherwise, without express written permission of the publisher.

ISBN: 9798322538196

Cover design by: EnQi ReaL

Printed in the United States of America

to & for God.

CONTENTS

Title Page
Copyright
Dedication
Diop 1
Electricity 101 8
Electricity 102 16
Human Engineering 20
Heart to Heart 24
Middle School 31
High School 93
Dollard Builds... 102
College 106
Microtubule Data 120
Post Grad 150
The Was Sceptre 188
Human Chemiluminescence of Phosphorus is the source of Human Bioluminescence and 192

Human Photoluminesc

DIOP

Diop was the GOAT of his time...

Ross Aiken Gortner's research was ignored or covered up... my thoughts on him anyway...

"The Egyptians, though healthy, large and robust were clumsy in their forms and coarse in their features. Like other African tribes they were woolly haired, flat-nosed and thick lipped, and IF NOT ABSOLUTELY BLACK WERE VERY NEAR IT IN COLOR."

Specimens of Ancient Sculpture

Aegyptian, Etruscan, Greek and Roman

Richard Payne Knight

British Archeologist / Historian

Source: -Specimens of Ancient Sculpture, Aegyptian, Etruscan, Greek, and Roman, Society of Dilettanti, Vol 1, London: by T. Bensley for T. Payne and J. White, 1809-1835. Published by Richard Payne Knight

"No people has bequeathed to us so many memorials as to their form, complexion and physiognomy as the Egyptians. If we were left to form an opinion of the subject by the description of the Egyptians left by the Greek writers we should conclude that they were, if not negro at least closely akin to the negro race. THAT THEY WERE MUCH DARKER THAN THE NEIGHBORING ASIATICS; that they had hair frizzled either by nature or art; that their lips were thick and projecting and their limbs slender, rest upon the authority of eye witnesses who had travelled in the country and who could have had no motive to deceive...the fullness of the lips seen in the sphinx of the pyramids and in the portraits of the kings is characteristic of the negro."

The Ancient History Of The East

Philip Smith

British Historian/Scholar

"During an excavation headed by the German Institute for Archaeology, Cairo, at the tombs of the nobles in Thebes-West, Upper Egypt, three types of tissues from different mummies were sampled to compare 13 well known rehydration methods for mummified tissue with three newly developed methods. .. Skin sections showed particularly good tissue preservation, although cellular outlines were never distinct. Although much of the epidermis had already separated from the dermis, the remaining epidermis often was preserved well (Fig. 1). THE BASAL EPITHELIAL CELLS WERE PACKED WITH MELANIN AS EXPECTED FOR SPECIMENS OF NEGROID ORIGIN."

--(A-M Mekota and M Vermehren. (2005) Determination of optimal rehydration, fixation and staining methods for histological and immunohistochemical analysis of mummified soft tissues. Biotechnic & Histochemistry 2005, Vol. 80, No. 1, Pages 7-13

In practice it is possible to determine directly the skin color and hence the
ethnic affiliations of the ancient Egyptians by microscopic analysis in the
laboratory; I doubt if the sagacity of the researchers who have studied the
question has overlooked the possibility.

Melanin (eumelanin), the chemical body responsible for skin pigmentation, is,
broadly speaking, insoluble and is preserved for millions of years in the skins
of fossil animals.

20 There is thus all the more reason for it to be readily recoverable in the

skins of Egyptian mummies, despite a tenacious legend that the skin of mummies,
tainted by the embalming material, is no longer susceptible of any analysis.

21 Although the epidermis is the main site of the melanin, the melanocytes
penetrating the derm at the boundary between it and the epidermis, even where
the latter has mostly been destroyed by the embalming materials, show a melanin
level which is non-existent in the white-skinned races.

The samples I myself analyzed were taken in the physical anthropology
laboratory of the Mus'ee de l'Homme in Paris off the mummies from the Marietta
excavations in Egypt.

22 The same method is perfectly suitable for use on the royal mummies of
Thutmoses III, Seti I and Ramses II in the Cairo Museum, which are in an excel
state of preservation.

For two years past I have been vainly begging the curator of the Cairo Museum
for similar samples to analyze. No more than a few square millimetres of skin
would be required to mount a specimen, the preparations being a few um in
thickness and lightened with ethyl benzoate.

They can be studied by natural light or with ultra-violet lighting which
renders the melanin grains fluorescent.

Either way let us simply say that the evaluation of melanin level by
microscopic examination is a laboratory method which enables us to classify the
ancient Egyptians unquestionably among the black races.

By Cheikh Anta Diop

Gloger's Law which states "Warm-blooded beings are pigmented in hot and humid climates."

Gloger's rule is an ecogeographical rule which states that within a species of endotherms, more heavily pigmented forms tend to be found in more humid environments, e.g. near the equator. It was named after the zoologist Constantin Wilhelm Lambert Gloger, who first remarked upon this phenomenon in 1833 in a review of covariation of climate and avian plumage color. [1] Erwin Stresemann later noted that the idea had been expressed even earlier by Peter Simon Pallas in Zoographia Rosso-Asiatica (1811). [2] Gloger found that birds in more humid habitats tended to be darker than their relatives from regions with higher aridity. Over 90% of 52 North American bird species studies conform to this rule.[3]

One explanation of Gloger's rule in the case of birds appears to be the increased resistance of dark feathers to feather- or hair-degrading **bacteria** such as Bacillus licheniformis. Feathers in **humid environments have a greater bacterial load**, and **humid environments are more suitable for microbial growth**; dark feathers or hair are more difficult to break down.[4] More resilient eumelanins (dark brown to black) are deposited in hot and humid regions, whereas in arid regions, pheomelanins (reddish to sandy color) predominate due to the **benefit of crypsis**.

Among mammals, there is a marked tendency in equatorial and tropical regions to have a darker skin color than poleward relatives. In this case, the underlying cause is probably the need to better protect against the more intense solar UV radiation at lower latitudes. However, absorption of a certain amount of UV radiation is necessary for the production of certain vitamins, notably vitamin D (see also osteomalacia).

Gloger's rule is also vividly demonstrated among human populations.[5] Populations that evolved in sunnier environments closer to the equator tend to be darker-pigmented than populations originating farther from the equator. There are exceptions, however; among the most well known are the Tibetans and Inuit, who have darker skin than might be expected from their native latitudes. In the first case, this is apparently an adaptation to the extremely high UV radiation on the Tibetan Plateau, whereas in the second case, the necessity to absorb UV radiation is alleviated by the Inuit's diet, which is naturally rich in **vitamin D**.

Crypsis is the ability of an animal or a plant[1] to avoid observation or detection by other animals. It may be a **predation strategy** or an **antipredator adaptation**. Methods include **camouflage, nocturnality, subterranean li festyle** and **mimicry**. Crypsis can involve visual, olfactory (with **pheromones**) or auditory concealment. When it is visual, the term cryptic coloration, effectively a synonym for animal camouflage, is sometimes used, but many different **methods of camouflage** are employed by animals or plants.

ELECTRICITY 101

The human body does not run on Electrons, nor does it run on electricity, it runs on light! I have been Hated by Dr. Sebi and Robert Becker followers as well as PLAGIARIZED BY THE LIKES OF ANDREW HUBERMAN, **JACK KRUSE** & LYOR COHEN...

No man can close the door GOD has opened for you! That being said we can start with Neurons, Nerves & Muscles. Keep in mind that Neurons & Melanocytes are fraternal twins. We are going to engage in a huge futuristic thought experiment.

What if the body was Electric? Like sold in Best Buy type.... What can we learn?

These cells are like unique magnets, very complex in their Russian doll structure. The core thing that keeps everything in the Gel ordered is Male/Female relationship. If that gets lost the cells will Die. When I say male/female I am speaking to the positive/negative charge aspect, polarization.

The only difference between a cell and a

traditional bar magnet, is that the cell is a blob of gel or thick fluid, the bar magnet is 'solid' and linear. A thin oblong square of steel or iron, where 1 side is 'north' and the other side is 'south'. Cells like I have been telling you for over 20 years now are built like Russian dolls. This means instead of one side vs the other side, its inside vs the outside. For energy to flow unlike magnets, this polarization is able to change. The charge values oscillate. The numbers aren't even that important in the grand scheme of things... I know you like "What numbers not that...WHAT"...LOL

The polarization of ElectroMagnetic cells, is negative charge on the inside, with a more positive charge on the outside. I didn't say positive, I said more positive. It's about the ratio.

This is why we need to discuss ions ad nauseam in the previous books, its the collections of ions that determine the charge values. Ions carry the charges in and out of the cells. If this becomes clear to you, you will see the basic binary nature of life, duality. Genes are switched off or on, it's not more complicated than that. Negative ions are electron driven and Positive ions are proton driven. Atoms combine to form ions.

The negative charge, or more negative charge, on the inside vs the outside, is created by a uneven distribution of ions. This uneven distribution creates the electric potential. This electric potential can be destroyed by to much bad water at once, water intoxication. This is why we add monatomic to every bottle, plastic container, jug, cup or mug of water we drink. If we drink from a spring or straight off a pipe (pause) then we take the Momatomix directly under our tongue.

Movement of the ions, across the cell membrane creates electricity or in biochemistry, a action potential. At the end of the day every cell works this way to some degree. We won't add the complexity of microtubules and organelles just yet, its more important to get the basics 1st.

*The beauty of these books that we create is I know you can feel yourself growing more cerebral, more intelligent. That is literally brain development and neural network growth! We are building Neural Plasticity in ReaL Time (triple entendre)!

This is why we had to get those last couple books out the way, You got some Nerve, Mitochondria Water & the Fascia Book... This book in your hands wouldn't full click without those being read and those won't fully clicked with the previous books etc... We are building!

Movement of the ions, across the cell membrane creates electricity or in biochemistry, a action potential. At the end of the day every cell works this way to some degree. Nerves communicate with each other and all other cells this way, the cells of your glands produce hormones this way etc... Your heart beats this way, the movement of Atoms from inside cells to outside cells and back.

Your cell may go from a resting charge of -70mV to its action state of +40mV and then back to -70mV. There have been authors that attributed all sickness to the malfunction of this basic system, many illnesses are but not all. Lets keep some things in mind before we go any further:

Electronics is a scientific and engineering discipline that studies and applies the principles of physics to design, create, and operate devices that manipulate electrons and other electrically charged particles. Electronics is a subfield of electrical engineering which uses active devices such as transistors, diodes, and integrated circuits to control and amplify the flow of electric current and to convert it from one form to another, such as from alternating current (AC) to direct current (DC) or from analog signals to digital signals. - wiki

God (the Sun) speaks (emits radio waves) that you hear (absorb and convert) through your antenna (the spine). The nerves help distribute this information to the Plasma based Crystal disc, fitted with integrated circuits as well as gates and channels.

The trillions of Computer Chips you have in your Electronic Body are wired by these nerves. How is it possible that the spine is known by Kemetic

Scientist and not it's make up? The spine is the Motherboard, I almost labeled it a backplane.

Your most potent antioxidants are semi-conductors. Electricity is movement and we interpret this as sound so we say God is Music.

Sonoluminescence & Chemiluminescence is our essence.

In each cell you have Mitochondria and Melanosomes, Mitochondria use Heme and Melanosome use Melanin. These two pigments allow their organelles to function like Engines/Motors and Alternators.

Engines and Motors - convert energy (mostly electric) into mechanical energy. Think electrons are converted into the movement of hydrogen ions, which then power the movement of the F0 motor, then recycling ATP, which releases the needed light we feed on only when a phosphate is oxides and released (movement on movement).

Alternators - convert mechanical energy into electrical energy. The movement of water creates electron pools that melanin harnesses, as well as the ROS, RCS and RNS (free radicals) released by the mitochondria. Melanin takes that energy created by movement and recycles it with NAD.

They aren't perfect analogues but its a great way to keep their activity in your memory.

Your brain is a processing unit, your heart is a turbine and hard drive.

Technology Equipment Recap...

Melanocytes - Photovoltaic Cells

Neurons/Nerves - Electromagnetic Cells

Fascia - Plasma Medium (misonomered Ether)

Brain - CPU, Inductor

Operating System - Deductive Logic or PQ

Heart - Turbine and Hard Drive.

Melanosomes - Alternators

Mitochondria - Motors

Cytoskeleton - Filaments

Phospholipids - Capacitors

Spine - Piezoelectric, Motherboard, Radio Wave Antenna

RBC - Floppy Discs

Protein - Transformer

Antioxidants - Semi-conductors (especially the selenium based…)

Body Cells - Plasma based Crystal disc, fitted with integrated circuits as well as gates and channels

Nerves & Vessels - 'Copper' wires and Fiber Optics

Pigment, Nerve & Blood Clusters - Input Devices like a Mouse, Keyboard, etc...

DNA - Piezoelectric, Antenna, Data Storing Inductors.

Stomach - Massive Inductor

Collagen Based Tissue - Piezoelectric Inductors

Lumen - the SI unit of luminous flux = to the amount of light emitted per second..... or hollow structures in vessels and cells... hmmm????

Eyes - Camera Lens/Charge Coupled Device (CCD), Digital to Analogue Converter, Complex Photovoltaic Cells/Photodetector...

Plasma based Crystal disc, fitted with integrated circuits as well as gates and channels fitted in a Capacitor (membrane). The full body capacitors are also fitted with a wide variety of antenna and transformers. These Nanobots have 2000+ alternators and motors in them for internal power. Each of these cells are Electrochemical with thousands of smaller nano electric machines.

Do you see how the body's technology is stacked or how the body as technology is stacked? Having DNA makes each cells it's own CPU, this could be the real reason the brain is so hard to figure out. We know the brain and mind are separate but....

The brain may be more of a Server than a CPU. If each cell is its own CPU, and the brain seen more as a processor for all the processors, it makes more sense. Food for thought...

ELECTRICITY 102

Lets stay with the Neurons, -70 at rest and +40 at work. In Ancient Egypt they recognized very early that Sodium or Salt had a special relationship with the human body. Water and Electricity are lead by Sodium. Sodium and it's partner in crime Potassium.

Potassium on the inside of the cell body potentiates the -70mV charge and when the gates of the cell are opened sodium is funneled in and potassium is released. Sodium makes the cell jump to action, as the inside of the cell with all the sodium is now +40 compared to the outside of the cell.

*Its -70mV compared to the outside of the cell at rest and +40mV compared to the outside of the cell at work. Review times table question and answer #10.

Let me apologize in advance, you will definitely need at a bare minimum, to have read the Fiscal Edition, Algarhythm & L'Goat books. I am realizing there is to much explaining I would have to backtrack in this book and my aim is simplicity.

Readers are Leaders!

Pages 12-14 in Melanin vs Diabetes Book One we discuss the nature of your Operating System as Deductive Logic. Conditional Arguments, If-Then Arguments, Hypothetical Syllogisms are the origin of the slang term Ps & Qs. Mind your Ps and Qs. This is yet another proof that Freemasonry is based on biochemistry! The 7 liberal arts that form the basis of Masonic Education is grounded in the Artform of Critical Thinking & Deductive Reasoning. This is probably why I was hooked as a kid on Encyclopedia Brown, those books combined Logic and Science...

The cell is at rest at -70mV, this means that the water is structuring properly, the nDNA is functioning properly, the Mitochondria etc... -70mV is where everything in the cell has just the right amount of power to work. That means when the channels, gates or porins open and change that value, the cell's response is to immediately change it back.

The same process then happens in reverse, the sodium is pushed out of the cell and the potassium is pulled back into the cell.

That is a pulse or oscillation.

That is a 1 or 0.

Do you see what's happening?
Now how can this happen Quadrillions of times per second, without any potassium?

That's called Hypertension.

That is why the model operandi is stay away from Salt, especially if your Black! See?? Health always comes down to Black or White, Equatorial vs Polar...

EuMelanin concentrates sodium, the reason is again, our Haplotypes are based around a lush green environment with lots of C3 food. That type of diet is a 7 to 1 ratio of Potassium to Sodium, the diet we consume now is 7 to 1 Sodium to Potassium, we are eating, living backwards which is evil lol...

Neurons generate signals in their cell bodies, their axons either connect to another Neuron/Nerve, a Muscle Cell (Myocyte) or a Gland for the desired response. Desired being the keyword here. In the Movement Book on pages 23-27 we begin the discussion on this relay-tionship. In truth we began in Pages 12-14 of Melanin vs Diabetes Book One discussing the nature of your Operating System as Deductive Logic.

How experience becomes information and your personality is created. We then spent the greater part of the Movement, AlgaRhythm & Divine Mathematics books showing you how to change and/or upgrade your Operating System. In fact we began with Melanin vs Diabetes Book two & three...

Point being the body is a complex system but it's operations aren't so complicated, it was designed for a baby to operate literally.

Physical Damage... Block signal generation, signal targeting, signal reception, input, processing...

Bad Nutrition... changes the resource pool for reactions, reaction speeds, repair... or inhibits them all together!

Toxins & Viruses... change programs or act as malware... changing the way your system operates or changing the way specific parts function...

Bacteria - can harbor or create toxins &/or viruses...

Epilepsy - sudden blast of electrical activity or too much electricity generate for a thought or action.

MS - lack of insulation on the wiring

Alzheimers, Dementia & Parkinson's - the very opposite of Epilepsy, not enough electrical activity &/or loss of neurons all together...

HUMAN ENGINEERING

Electrical Circuit - a closed loop, electrical network, that has a return path for its current. This network must contain active electronic components that can compute, transfer data and perform signal, power or data amplification.

Circuit Training is actually the best form of workout outside of Isometrics and Calisthenics. Circuit Training with blended aspects of High Intensity Interval Training! Especially for Fast Twitch Dominant folks... This should be becoming obvious to you now...

To simplify what circuit training is, you don't take any breaks. You move directly from one set to the next and after you do all the exercises you have in your routine you may rest 1-3 minutes and then you begin again!!! It infuses Cardio into your program very masterfully and if you increase your body heat with a hoody your at a goat level!!! Just make sure your hydrating and adding the appropriate levels of Momatomix!

Basic Electronic Circuit Components

Resistor - a resistor slows down current or signals.

Transistor - the opposite of a resistor, transistors are **semiconductors** that control and amplify signals or current.

Lumen - the SI unit of luminous flux = to the amount of light emitted per second….. or hollow structures in vessels and cells… hmmm????

Filament - a slender threadlike object or fiber, especially one found in animal or plant structures or a conducting wire or thread with a high melting point, forming part of an electric bulb or vacuum tube and heated or made incandescent by an electric current.

Transformer - transfers one type of energy to another type, increasing (stepping up) or reducing (stepping down) the voltage. Transformers can also share energy from one circuit to one or more others.

Capacitors (Condenser) - a capacitor stores charge, using two parallel or close surfaces, insulated from one another.

Inductor/Reactor - (this is a very popular design in Human Engineering) a coil that stores energy in a magnetic field as electricity runs through it, usually consists of coiled insulated wire.

Diode - a semiconductor with two (di) terminals that allows current or signals to flow in only one direction. A thermionic tube having two electrodes (a anode and cathode). Keep in mind the basic design of the spine and its Radio Wave nature, nerve supply, it's connection to the Aorta, it's active Marrow & its **Piezoelectric**. The spine is a real work of art... State of the Art...

Anode - the positively charged electrode by which electrons leave a device or the negatively charged electrode of a device supplying current such as a primary cell.

Cathode - the negatively charged electrode by which electrons enter a device or the positively charged electrode of a device supplying current such as a primary cell.

*Notice that all of the Nobel Prize Winners base theories of Health & Wellness on oxygen supply to the tissues, its because oxygen brings electricity, magnetism & **light with it**.

LET THERE BE LIGHT? LETS START HERE.... WITH HA TORAH

AWR OR AVR OR AUR (pronounce like oar or Aura) - LIGHT

AR - ROOT OF LIGHT MEANS ORDER OR BOX

Is Hebrew or the creator of Hebrew saying Light come from a Box?

HEART TO HEART

What can we learn from Electrical Heart Trouble?

In the Osiris, Diabetes & Respiration book pages 132 - 146 we discuss the issues surrounding Heart Attack, we have a whole book dedicated to religious, medical and holistic/naturopathic aspects of the Heart, complete w/ cheat codes on pages 60-63 in the Healthy Heart Book.

Here we want to do something different because honestly we have discussed Heart Health directly or indirectly for almost 30 books straight LOL...

We need to keep in our minds now that in our Hearts are regulated by Cardiac Melanocytes!

The Mitral, Tricuspid & Aortic 'Mouth' are loaded with Melanocytes.

The Atrial & Ventricular Septa are loaded with Melanocytes.

The Pulmonary Vessels are also all loaded with Melanocytes.

The density in the heart and pulmonary vessels, is

reflected by the density in the skin. In the world where being dark skinned like me means higher cancer risks, metabolic disease risk etc... Its great to know that this one thing is on my side lol... This also means that it could be the hundreds of years of Medical Apartheid and Institutionalized Racism has a big role to play in heart disease! Academia is largely ignoring the Melanocytes in these locations!!! If you don't understand what I am saying then let me add a little information.

All 4 Chambers of the Heart are Pigmented!!!

All of the Valves of the Heart are Pigmented!!!

It stands to reason that **Heart Health = Melanocyte Health**!!!

Now do you see what I have been saying?

Did you peak through those bars and see the hidden discovery? Oh yeah Jack Kruse and all his minions will be thirsty to steal this LMMFAO!!!!

The density and activity of the internal Melanocytes matches the skin! That means what? This is the secret benefit of the sun!

This is the BIG ONE!!!
The sun affects the Lungs and Heart by regulating the Melanocytes in the Lungs and Heart!!! Yes you heard it here first... Guess what the whole world can benefit from this too... Unlike Vitamin D, you only need UVA for this, just go outside anytime its

sunny as naked as possible!

Direct Sunlight on the neck and chest is preferred!!!

Cardiac Arrest the strangest of all Heart Afflications, with its less demons, variants of arrhythmia, tachycardia and bradycardia.

Arrhythmia - Irregular Beat (to fast or too slow or off beat).

Tachycardia - Fast heart beat (sympathetic).

Bradycardia - Slow heart beat (parasympathetic).

The first thing we should notice is balance is required from your heartbeat, not too fast or too slow.

Exercise is a easy first line of defense for Bradycardia, in our society that is not the big issue unless you have a drug problem. Then we still need to address the drugs before the heart.

The only complication is the people who aren't in extremely good shape, need to consult their doctor on what type of exercise would be safe, over working yourself could be just as dangerous as the condition itself.

Do not confuse a low Resting Heart Rate with Bradycardia, the better shape you are in the lower you Resting Heart Rate will go and the Higher your Basal Metabolic Rate. Those are goals! Don't be scared or confused, but if your in not so good shape and your ticker isn't ticking...

The more we Eat Right for our Haplotypes and exercise, we get stronger Parasympathetic Nervous Systems, which fight responses to Adrenalins/Fight/Flight and stress chemicals. This means that being in bad shape means being highly sensitive to stress! Nervous, Paranoid and Defensiveness are all states of mind, we are showing are clearly attached to Heart Function & Cardiovascular Health! Just refer to the Ebers Papyrus section of the Healthy Heart Book!

The **biggest** impact on Healthy People having lower RHRs is what we uncovered in the Aorta books! The skeletal muscle is apart of the Circulatory System and the Digestive System. The more skeletal muscle you have the less you need the heart to pump the blood for you, and the more often you work those 600-800 muscles the less pressure is on the Heart!!! Naturopathic and Holistic Care is about prevention.

*Bradycardia Keys

Get the Body Fat Down (use Melanin vs Diabetes the Fiscal Edition, the Gold Book, Divine Mathematics Book & AlgaRhythm Book)

Get walking! Add in standing heel raises and toe raises daily!

Stop Smoking Period!

Stop Drinking Alcohol & eating Yeast!

Stop the Sweetness and Sugary Snacks!

Now lets get on over to the too much sauce side of things.

The Valsalva Maneuvers mimic straining to poop or blow up a small tight balloon.

Step 1) The Valsalva Maneuvers is the first line of defense for many fast heart rate situations. You simply clench your abs, by attempting to breath out as hard as you can with your nose and mouth closed. In many instances this will slow down your heart beat.

This should add much more seriousness to 'cleaning your apron' or working out your abs! The diaphragm sits just under the lungs and heart, it is literally a pumping muscle. The pumping of the diaphragm is responsible for respiration, venous circulation and to some extent the lymph!

The diaphragm muscle extends from the bottom of the Mouth to the Pelvic Floor! This is the reason why we have the pictures in the Clean Abs Fascia Book! You gotta see this stuff! This means all of your abs are directly or indirectly tied (literally) to your diaphragm! That's how important the Heart and Lungs are, the site of the most important Melanocytes!

This is why the Heart and its 12 divisions was called the Throne of Osiris! Osiris is buried in your Chest! Your blood is made of Sweet Wine or Glucose, you are the people who die after these big meals!!!

Do you see all we are learning by looking at the steps to heal electrical malfunctions in the Heart?

Step 2) Adenosine injections. Yep, the same adenosine from your DNA and ATP. Remember adenosine is apart of the sleepy time crew. When the mitochondria can't recycle the ATP fast enough, the adenosine acts a vasodialator to signal the pineal to produce melatonin… That Melatonin tells the body's cells to produce they own melatonin and soon we off to counting sheep…

In theory a nice pure shot of that should tranquilize your heart, sometimes it does too. It does fail sometimes as well and then the pull out the electric pushups!

Step 3) Clear! Pooofff! Pushups have the ability reduce the risk of Cardiovascular Disease by 97% if you can do more than 40 in 60 seconds! This is simple because of all we have share in the Clean Abs and You got some Nerve books, you should able to begin to put these together, your are electrically stimulating the Heart!

You are building the muscles and cleaning the fascia closest to the Heart!

This is why we promote aiming for 60 pushups in 60 seconds shirt off in the sun!!!

Obviously if you have some health conditions this may not be right for you but man....

When it come to a once of prevention being worth a pound of cure!!! Stop it! There is no drug in existence offering a documented 97% reduced risk of cardiovascular disease and now you know both reasons why! The sun is a doctor! Sunlight speaks to your spine and provides solar nutrition to your Melanocytes!

Those Melanocytes translate that into their language which we have been teaching you since pages 39-42 in the Fiscal Edition book and pages 39-40 in L'Goat book! That's literally the secret language they use to control inflammation and turn lemons (bad cytokines) into lemonade (energy).

MIDDLE SCHOOL

Suntans. Between 10am & 2pm is UVB time, that means your getting UVA & UVB. This is the time of day when the sun produces melanin (melanocytes), vitamin d (cholesterol) and nitric oxide (arginine/citruline via heme) in you.

The Magic here is for those Cardiac & Pulmonary Melanocytes, depending on your skin tone, will determine how long you need to be in the Sun. Keep in mind that it may take up to 36 hours for the tan to become fully visible and that it may start to fade after a additional 36 hours. If you want to build your tan and permanently increase your melanin you need to get as much sun as is safe for your complexion and follow the endarkenment protocol on pages 7-9 in the Healthy Heart book.

You don't have to soak the midday sun you can rock out from 6am-9:30am and then from 3pm until...

Selenium the Semiconductor boost your body's tanning ability and so does the Batana Vitamins A&E!!!

Anyway I feel like I have been fighting such a uphill battle to get my first discovery out there.... You know that **Melanin is Photovoltaic**! That we haven't went to the in between. We discussed Melanin being Photovoltaic and that because of our many uses of phosphorus the cell itself is Chemiluminescent. Here is what we missed, Electroluminescence. All photovoltaic cells are also electroluminescent, its basically the proofing.

Electrician test photovoltaic cells by running them in reverse to see if they light up, if they light up, they are in good working condition and if they don't they need repair. Melanocytes just like Neurons, their fraternal twin, can theoretically produce light!

In electromagnetism, current density is the amount of charge per unit time that flows through a unit area of a chosen cross section. [1] The current density vector is defined as a vector whose magnitude is the electric current per cross-sectional area at a given point in space, its direction being that of the motion of the positive charges at this point. In SI base units, the electric current density is measured in amperes per square metre.

In physics, electromagnetism is an interaction that occurs between particles with electric charge via electromagnetic fields. The electromagnetic force is one of the four fundamental forces of nature. It is the dominant force in the interactions of atoms and molecules. Electromagnetism can be thought of as a combination of electrostatics and magnetism, which are distinct but closely intertwined phenomena.

Electrostatics is a branch of physics that studies slow-moving or stationary electric charges.
Since classical times, it has been known that some materials, such as amber, attract lightweight particles after rubbing. The Greek word for amber, ἤλεκτρον (élektron), was thus the source of the word 'electricity'. Electrostatic phenomena arise from the forces that electric charges exert on each other. Such forces are described by Coulomb's law.

Coulomb's inverse-square law, or simply Coulomb's law, is an experimental law[1] of physics that calculates the amount of force between two electrically charged particles at rest. This electric force is conventionally called the electrostatic force or Coulomb force.

Electric charge (symbol q, sometimes Q) is the physical property of matter that causes it to experience a force when placed in an electromagnetic field. Electric charge can be positive or negative. Like charges repel each other and unlike charges attract each other.

An electric current is a flow of charged particles, [1][2][3] such as electrons or ions, moving through an electrical conductor or space. It is defined as the net rate of flow of electric charge through a surface.[4]:2[5]:622 The moving particles are called charge carriers, which may be one of several types of particles, depending on the conductor. In electric circuits the charge carriers are often electrons moving through a wire. In semiconductors they can be electrons or holes. In an electrolyte the charge carriers are ions, while in plasma, an ionized gas, they are ions and electrons.[6]

In the International System of Units (SI), electric current is expressed in units of ampere (sometimes called an "amp", symbol A), which is equivalent to one coulomb per second. The ampere is an SI base unit and electric current is a base quantity in the International System of Quantities (ISQ).[7]:15 Electric current is also known as amperage and is measured using a device called an ammeter.[5]:788

Electric currents create magnetic forces, which are used in motors, generators, inductors, and transformers.[8][9] In ordinary conductors, they cause Joule heating, which creates light in incandescent light bulbs. Time-varying currents emit electromagnetic waves, which are used in telecommunications to broadcast information.

The joule (pronounced /ˈdʒuːl/, JOOL or /ˈdʒaʊl/ JOWL; symbol: J) is the unit of energy in the International System of Units (SI).[1] It is equal to the amount of work done when a force of one newton displaces a mass through a distance of one metre in the direction of that force. It is also the energy dissipated as heat when an electric current of one ampere passes through a resistance of one ohm for one second. It is named after the English physicist James Prescott Joule (1818–1889).

The watt (symbol: W) is the unit of power or radiant flux in the International System of Units (SI), equal to 1 joule per second or 1 kg·m2·s-3. [1][2][3] It is used to quantify the rate of energy transfer. The watt is named in honor of James Watt (1736–1819), an 18th-century Scottish inventor, mechanical engineer, and chemist who improved the Newcomen engine with his own steam engine in 1776. Watt's invention was fundamental for the Industrial Revolution.

The ampere (/ˈæmpɛər/ AM-pair, US: /ˈæmpɪər/ AM-peer;[1][2][3] symbol: A),[4] often shortened to amp,[5] is the unit of electric current in the International System of Units (SI). One ampere is equal to 1 coulomb (C) moving past a point per second.[6][7][8] It is named after French mathematician and physicist André-Marie Ampère (1775–1836), considered the father of electromagnetism along with Danish physicist

Hans Christian Ørsted.

As of the 2019 redefinition of the SI base units, the ampere is defined by fixing the elementary charge e to be exactly 1.602176634×10-19 C,[6][9] which means an ampere is an electric current equivalent to 1019 elementary charges moving every 1.602176634 seconds or 6.241509074×1018 elementary charges moving in a second. Prior to the redefinition the ampere was defined as the current passing through two parallel wires 1 metre apart that produces a magnetic force of 2×10-7 newtons per metre.

The earlier CGS system has two units of current, one structured similarly to the SI's and the other using Coulomb's law as a fundamental relationship, with the CGS unit of charge defined by measuring the force between two charged metal plates. The CGS unit of current is then defined as one unit of charge per second.

The volt (symbol: V) is the unit of electric potential, electric potential difference (voltage), and electromotive force in the International System of Units (SI).

One volt is defined as the electric potential between two points of a conducting wire when an electric current of one ampere dissipates one watt of power between those points.

In electromagnetism and electronics,

electromotive force (also electromotance, abbreviated emf,[1][2] denoted or [citation needed]) is an energy transfer to an electric circuit per unit of electric charge, measured in volts. Devices called electrical transducers provide an emf[3] by converting other forms of energy into electrical energy.[3] Other electrical equipment also produce an emf, such as batteries, which convert chemical energy, and generators, which convert mechanical energy.[4] This energy conversion is achieved by physical forces applying physical work on electric charges. However, electromotive force itself is not a physical force,[5] and ISO/IEC standards have deprecated the term in favor of source voltage or source tension instead (denoted).

In the context of the gyrator-capacitor model of a magnetic circuit, magnetic inductance (SI unit: F) is the analogy to inductance in an electrical circuit.

In electromagnetism, a dielectric (or dielectric medium) is an electrical insulator that can be polarised by an applied electric field. When a dielectric material is placed in an electric field, electric charges do not flow through the material as they do in an electrical conductor, because they have no loosely bound, or free, electrons that may drift through the material, but instead they shift, only slightly, from their average equilibrium positions, causing dielectric polarisation. Because of dielectric polarisation, positive charges are displaced in the direction of the field and negative charges shift in the direction opposite to the field. This creates an internal electric field that reduces the overall field within the dielectric itself. If a dielectric is composed of weakly bonded molecules, those molecules not only become polarised, but also reorient so that their symmetry axes align to the field.[1]

The study of dielectric properties concerns storage and dissipation of electric and magnetic energy in materials.[2][3][4] Dielectrics are important for explaining various phenomena in electronics, optics, solid-state physics and cell biophysics.

The triboelectric effect (also known as triboelectricity, triboelectric charging, triboelectrification, or tribocharging) describes electric charge transfer between two objects when they contact or slide against each other. It can occur with different materials, such as the sole of a shoe on a carpet, or between two pieces of the same material. It is ubiquitous, and occurs with differing amounts of charge transfer (tribocharge) for all solid materials. There is evidence that tribocharging can occur between combinations of solids, liquids and gases, for instance liquid flowing in a solid tube or an aircraft flying through air.

Often static electricity is a consequence of the triboelectric effect when the charge stays on one or both of the objects and is not conducted away. The term triboelectricity has been used to refer to the field of study or the general phenomenon of the triboelectric effect,[1][2][3][4] or to the static electricity that results from it.[5][6] When there is no sliding, tribocharging is sometimes called contact electrification, and any static electricity generated is sometimes called contact electricity. The terms are often used interchangeably, and may be confused.

Triboelectric charge plays a major role in industries such as packaging of pharmaceutical powders,[3][7] and in many processes such as dust storms[8] and planetary formation.[9] It can also increase friction and adhesion. While many aspects of the triboelectric effect are now understood and extensively documented, significant disagreements remain in the current literature about the underlying details.

Whenever a solid is strained, electric fields can be generated. One process is due to linear strains, and is called **piezoelectricity**, the second depends upon how rapidly strains are changing with distance (derivative) and is called **flexoelectricity**. Both are established science, and can be both measured and calculated using **density functional theory** methods. Because flexoelectricity depends upon a gradient it can be much larger at the nanoscale during sliding or contact of asperity between two objects.[38]

There has been considerable work on the connection between piezoelectricity and triboelectricity.[79][80] While it can be important, piezoelectricity only occurs in the small number of materials which do not have inversion symmetry, [36] so it is not a general explanation. It has recently been suggested that flexoelectricity may be very important[61] in triboelectricity as it occurs in all insulators and semiconductors.[81] [82] Quite a few of the experimental results such as the effect of curvature can be explained by this approach, although full details have not as yet been determined.[62] There is also early work from Shaw and Hanstock,[33] and from the group of Daniel Lacks demonstrating that strain matters. [83][84][70]

A different type of model has been proposed by Robert Alicki and Alejandro Jenkins.[58] They argue that the electrons in the two materials that slide against each other have different velocities, and that quantum effects cause this imbalance to pump electrons from one material to the other.

- Electrostatic generator, machine to produce static electricity
- Electrostatic induction, separation of charges and polarization due to other charges
- Electrostriction, coupling between an electric field and volume of unit cells
- Electrohydrodynamics, coupling in liquids between electric fields and properties

- Flexoelectricity, polarization due to bending and other strain gradients
- Mechanoluminescence, light produced by mechanical action, often involving triboelectric effect
- Nanotribology, science of tribology (friction, lubrication and wear processes) at the nanoscale
- Piezoelectricity, polarization due to linear strains
- Polarization density, general description of the physics of polarization
- Static electricity, electric charge often but not always due to triboelectricity
- Tribology, science of friction, lubrication and wear
- Triboluminescence, light associated with sliding or contacts
- Work function, the energy to remove an electron from a surface

The triboelectricity of the Human Body

Author links open overlay panel
Renyun Zhang a, Magnus Hummelgård a, Jonas Örtegren a, Martin Olsen a, Henrik Andersson b, Ya Yang a c, Haiwu Zheng d, Håkan Olin a
Show more
Add to Mendeley
Share
Cite

https://doi.org/10.1016/j.nanoen.2021.106041
Get rights and content
Under a Creative Commons license
open access

Highlights

- Historical review of the studies relates to the human body's triboelectricity.

- A systematic review of the HBT in areas such as cosmetics, electrostatic discharge, and triboelectric nanogenerators.

- Review of the mechanisms of the hair and skin's triboelectrification.

Abstract

Triboelectrification (contact electrification) as a physical phenomenon appeared for the first time in a dialogue by Plato around 400 B.C. The phenomenon described in the dialogue is about amber that people wear attracting dry hair. The description also indicated that triboelectrification was first discovered on the human body. However, the studies that have been carried out on triboelectrification were mostly based on other materials, such as polymers, rather than on the human body. The invention of **triboelectric nanogenerators** (TENGs) has recently opened a door for both fundamental and applied research and brought triboelectrification into real applications. The human body's triboelectricity, as a vital part of studies, has also attracted much interest in the past ten years. Research and review articles were published during this period. However, few articles included the biological fundamentals of the triboelectrification of the human body. Moreover, most of the review articles missed two important parts: the **electrostatic discharge** (ESD) of the human body, which has been widely studied in electronics, and the cosmetics that reduce the triboelectrification of hair. A systematic review including the fundamentals and the applications could help readers understand the human body's triboelectricity. Given this, we proposed this review article on the human body's triboelectricity. The paper will cover a brief history and a brief mechanism summary of triboelectrification; the epidermis structure of the human hair and skin, including how the chemicals on the epidermal layer contribute to the skin's triboelectricity; fundamental studies of the human body's triboelectricity; and applications that utilize the human body's triboelectricity. Perspectives for future studies and conclusions will be given at the end of the review.

Graphical Abstract

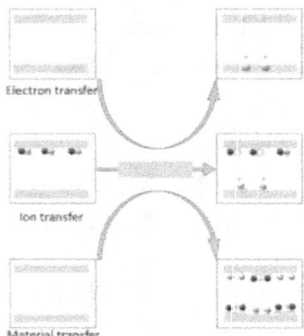

- Download : Download high-res image (180KB)
- Download : Download full-size image

- Previous article in issue
- Next article in issue

Keywords

TriboelectricityThe human bodyTriboelectric nanogeneratorsSensorsElectrostatic discharge

1. A brief history of triboelectrification

The first description of the triboelectric phenomenon, "the marvels that are observed about the attraction of amber" [1], appeared in Plato's dialogue Timaeus, which dates to approximately 400 B.C. The "attraction of amber" refers to the attraction of amber to dry hair that is a result of the electrostatic interaction in which electrostatic charges are generated by the triboelectrification between the amber and human skin or hair. Around 300 A.D., a Chinese Philosopher, Pu Guo [2], described the "amber effect" in his poem "Ci Shi Zan (Eulogy of the magnet).".

"磁石吸铁，琥珀取芥，气有潜通，数亦冥会，物之相感，出乎意外。".

Below is an English translation of the poem in Dr. Frank Nordhage's thesis [3]:

"The magnet draws the iron, and the amber attracts mustard seed. There is a breath which penetrates secretly and with velocity and which communicates itself imperceptibly to that which corresponds to it in the other object. It is an inexplicable thing."

"The amber attracts mustard seed" indicates the same physical principle as amber attracting dry hair. The "breath" describes the electric fields created by the triboelectric charges on the surface of the amber. The "which communicates itself imperceptibly to that which corresponds to it in the other object" explains how the charged object interacts with other objects.

William Gilbert experimentally proved [4] the generation of triboelectric charge by rubbing contact, written about in his book "de Magnete" in 1600. Such findings indicated the commonness of the generation of triboelectric charges between two surfaces in physical contact. The Swedish physicist J. C. Wilcke made the first triboelectric series [5] that showed the charge affinity of materials. The charge affinity indicates the charge transfer during the triboelectrification process. Fig. 1.

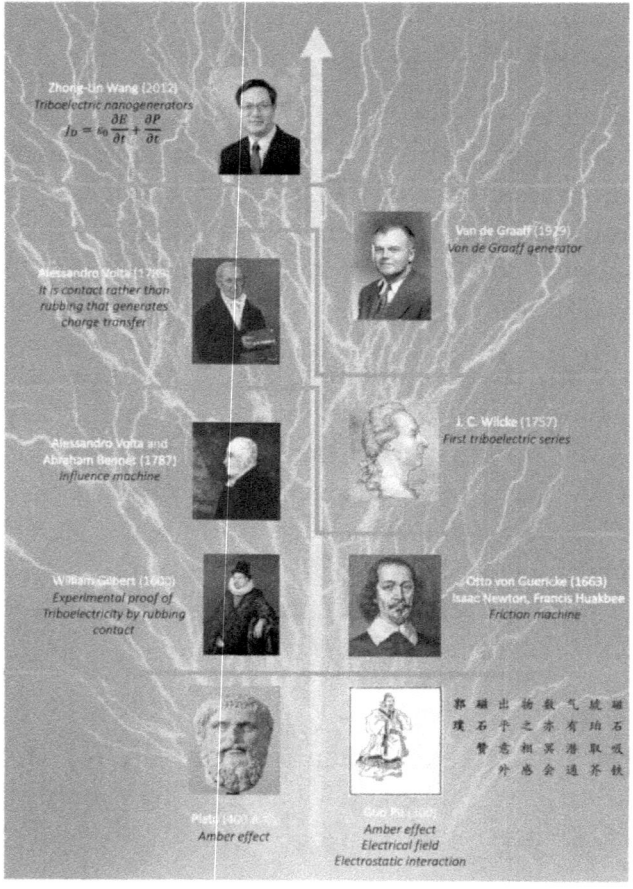

Download : Download high-res image (2MB)

Download : Download full-size image

Fig. 1. A brief history of triboelectrification.

All photos and pictures are copied from wikipedia.com.

During the triboelectrification process, the charge generation was recognized as being the result of rubbing two materials together for a very long time. It was not until 1789 when Alessandro Volta noted that it is contact, rather than rubbing, that leads to charge transfer; people started to realize that the actual mechanism was not that simple.

Despite the long history, the actual mechanisms of triboelectrification are still not fully understood. Different theories have been proposed, including ion transfer [6], **electron transfer** [7], and material exchange [8]. Recent experimental and theoretical approaches indicate that electron transfer dominates the charge transfer process of solid-solid interface [9].

Apart from theoretical studies, attempts to apply triboelectric charges have also been made for a very long time. The first attempt may be the friction machines that use the triboelectric effect. Otto von Guericke invented the first friction machine [10] around 1663, using the triboelectrification between hands and a sulfur globe. The sulfur globe was later changed to a glass globe, as suggested by Isaac Newton [11] and made by Francis Hauksbee [12]. The second attempt was the influence machine that uses electrostatic induction. Alessandro Volta [13] and Abraham Bennet [14] are the first two persons contributing to the invention. Despite these attempts, few applications have been used in practice. One of the known applications is the Van de Graaff generator [15], which has lately been used as a particle accelerator. Beside these machines that can use the human body's triboelectricity, there are other electrostatic generators such as the Kelvin water dropper [16], Holtz machine[17], and Wimshurst machine[18] that adopted other materials have been invented in the history.

A new chapter on triboelectrification started in 2012 when Zhong-Lin Wang and coworkers [19] invented a triboelectric nanogenerator (TENG). The TENG combined the triboelectric effect and electrostatic induction [20] to generate energy use in practice [21], [22], [23], [24], [25]. As the development has progressed [26], an increasing number of applications of TENGs have been discovered, such as sensors [27], actuation systems [28], control interfaces [29], [30], functional systems [31], and healthcare [32], [33]. Potential applications in the Internet of Things (IoT) [34] and artificial intelligence [2] will make TENGs have a high impact on modern information technologies.

2. A brief summary of the mechanisms of triboelectrification

Triboelectrification is one of the most complex processes that remain to be understood by scientists. Several mechanisms (Fig. 2) have been proposed in history, including ion transfer [35], electron transfer [36], and material transfer [37]. Debates on these three mechanisms have continued for many decades. Each mechanism has a range of applications due to the diversity of material choices in triboelectrification studies. The diversity of materials can create a diversity of contacts [38], such as metal-metal [39], metal-insulator [40], metal-polymer [41], polymer-polymer [42], inorganic-organic [43], [44], liquid-solid [45], [46], and liquid-liquid [47], [48]. The diversity of contacts implies that the charge transfer processes would be different for different interfaces.

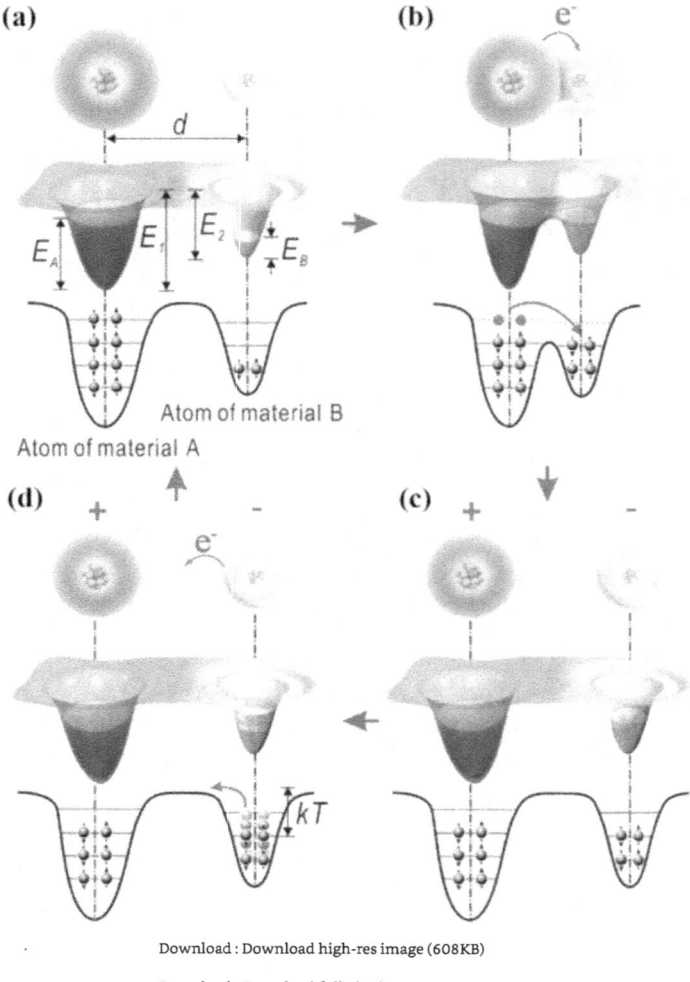

Fig. 2. An electron cloud–potential well model proposed for explaining CE and charge transfer and release between two materials that may not have a well-specified energy band structure, as in Fig. 5. Schematic of the electron clouds and

potential energy profiles (3D and 2D) of two atoms belonging to two materials A and B (a) before contact, (b) when they are in contact, and (c) after contact, showing electron transfer from one atom to the other after forcing the electron clouds to overlap. (d) Charge release from the atom at an elevated temperature T once kT approaches the barrier height. d, distance between two nuclei; EA and EB, occupied energy levels of electrons; E1 and E2, potential energies for electrons to escape; k, **Boltzmann constant**; and T, temperature.

Ref. [7] Copyright 2018, John Wiley & Sons, Ltd.

Researchers who prefer the ion transfer [49], [50] mechanism recognize that there are no **free** electrons in insulators. Therefore, electron transfer does not occur during the triboelectrification process. Experimental evidence of ion transfer came earliest from the printing industry, where charge control reagents were used [8]. Recently, a new concept of **mobile ions** [50], [51] was introduced to the theory, in which mobile ions can freely transfer between surfaces. However, the driving force of mobile ions has not been clearly explained [8]. Researchers also built a model based on a water layer [50] on the surface and explained the transfer of charges. However, such a charge transfer process can even occur in conditions without the presence of water [37].

The electron transfer mechanism [9] can be established following the theories in solid-state physics, such as the work function [52]. Such a mechanism can explain the contacts between materials with defined work functions. However, for organic materials, such as insulating polymers, the work function model does not work very well. An electron cloud-potential well model [53] has been recently proposed (Fig. 3) with the awareness that insulating polymers have no well-specified band structure. In this model, the contact of two atoms results in overlap of the electron clouds, leading to electron flow from a higher energy state to a lower energy state. A model based on quantum chemistry has been presented, in which the HOMO and LUMO orbitals [9] of the materials act similarly to the conduction and valence bands of semiconductors.

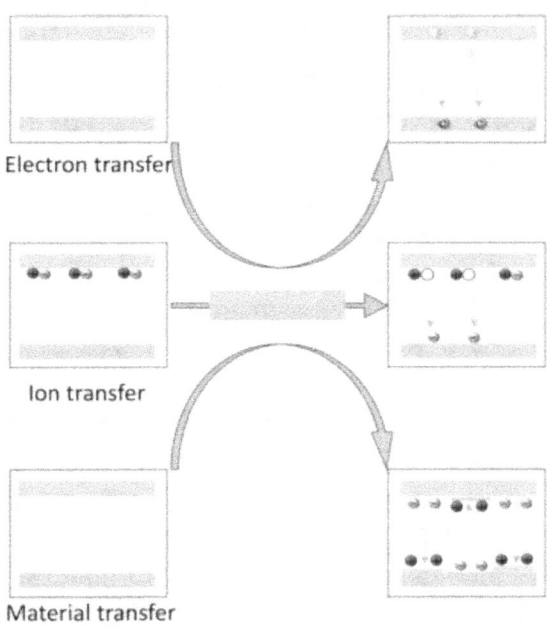

Fig. 3. Different mechanisms of triboelectrification.

Material transfer during the triboelectrification process has been experimentally proven using the XPS method [37]. Small clumps of materials transfer between surfaces during the contact-separation process. Such behavior implies broken **covalent bonds** on the polymer chain, creating free radicals that are chemically reactive and subsequently react with oxygen or water to form **charged species**.

Apart from the discussion of the mechanisms, a

new achievement that is of great important is the displacement current theory developed by Prof. Wang [54]. The theory adds a new term (∂/∂) to the Maxwell's equation to describe the current generated by a polarized field. Based on this theory, it is possible to quantitatively describe the current density of TENGs.

Generally, all previous experimental and theoretical studies have demonstrated that the mechanism of triboelectrification is very complicated. There is no general theoretical model that simply describes all of the charge transfer process. In addition, triboelectrification is not just a simple physical process but also involves chemical processes in some cases. Therefore, multidisciplinary approaches need to be conducted to explore the mechanisms.

3. Fundamental studies of the human body's triboelectricity

3.1. Triboelectric properties of the skin and hair

Triboelectrification of the human body occurs on either the skin [55] or hair [56] where the human body comes into physical contact with other objects. Both the skin and hair have high positive charge affinities compared to most of the materials that have been tested for their charge affinities, as shown in the triboelectric series. Few studies have explained why the skin and hair have similar charge affinities despite their different appearances. An answer to the question can be an obtained if one investigates the chemical components of the skin and hair.

The hair cuticle is mainly composed of keratin (Fig. 4) [58], which is also a large part of the epidermis of human skin [59]. This component in common makes them have similar triboelectric properties. If we examine keratin's chemical structure, we can find that the structure is very similar to that of nylon [60]. As a triboelectric material, nylon has a very high positive charge affinity, which has been evidenced experimentally. Therefore, the similar chemical structure on the hair cuticle and the epidermis of the skin makes them have a high positive charge affinity.

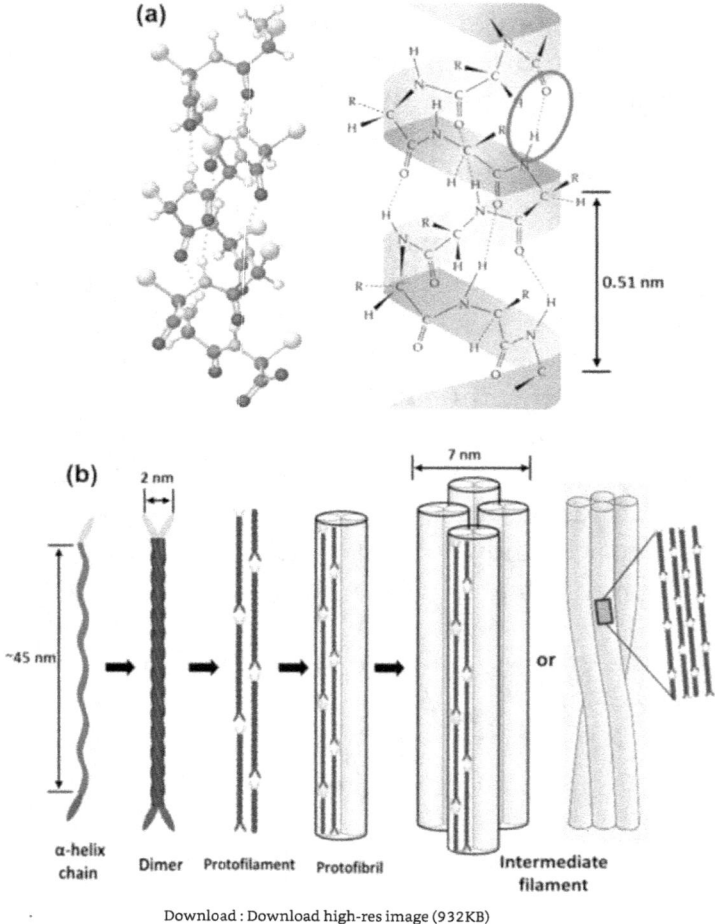

Fig. 4. Intermediate filament structure of α-keratin: ball-and-stick model of the polypeptide chain, and α-helix showing the locations of the **hydrogen bonds** (red ellipse) and the 0.51 nm pitch of the helix.

Ref. [57] Copyright 2016, Elsevier.

In some triboelectric series where both dry skin and hair are listed, the charge affinity of dry skin is higher than that of hair. The mechanism behind this is that the epidermal layer of the skin has more chemical components (**Fig. 5**) than hair. In a hair cuticle, almost 95% of the contents are keratin [62]. However, in the epidermis, the outermost layer (stratum corneum) contains not only keratin but also ceramides, cholesterol, and fatty acids [59]. The mixture of these chemicals may make the epidermis have a higher positive charge affinity than the hair cuticle. However, no direct evidence has been shown yet.

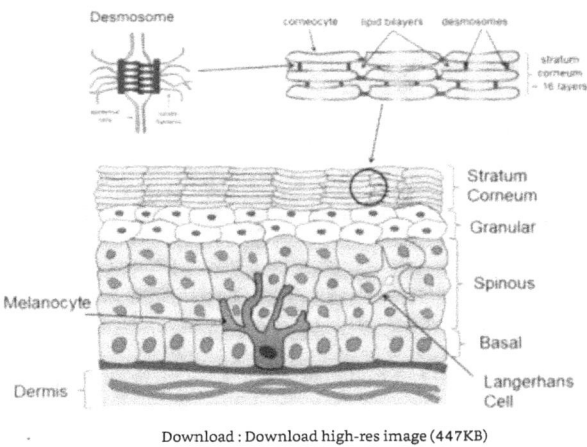

Download : Download high-res image (447KB)

Download : Download full-size image

Fig. 5. Skin structure. The components of the epidermal skin barrier are shown here. The outermost layer, the stratum corneum, is designed to be difficult to penetrate to protect from environmental insults. The granular, spinous and basal layers of the viable epidermis are responsible for generating and renewing the stratum corneum and are involved in wound healing. The epidermis also contains Langerhans cells and melanocytes. The skin barrier provides innate immune functions.

Ref. [61] Copyright 2014, Elsevier.

3.2. Triboelectric charge generation and transfer in the human body

3.2.1. Triboelectric charge generation

The charge generation process on human hair (Fig. 6) is similar to that for polymers commonly used in TENG studies. Electron transfer could be the primary process when hair is experiencing triboelectrification. Charge generation on hair has been tested by combing hair with different materials, such as metals and polymers. The charge density on hair depends on the materials used for rubbing as well as the direction of rubbing [63]. Modification of hair by reduction, bleaching, and oxidative dyeing has no significant effect on the hair's triboelectric effect [64]. However, modification of hair by a cationic surfactant or a cationic polymer could change its charge density [65]. If hair is immersed in acid or base solutions, then the charge transfer could also be changed, and it is believed that the mobile ions on the keratin surface of hair [66] contribute to the charge transfer. The use of an AFM-based Kelvin probe [67] led to a recent demonstration of the difference between virgin and damaged hair (chemically or mechanically) in charge generation, showing that charge generation on virgin hair is more stable than that on damaged hair with increasing load.

Download : Download high-res image (772KB)

Download : Download full-size image

Fig. 6. Frizzy hair and electrostatically charged hair when a kid is playing on a slide.

Photos are copied from Wikipedia.com.

Different from hair, charge transfer for the skin might be more complicated due to the complexity of the surface molecules. An argument can be made on which process dominates the charge transfer. The reason for making the argument is that material transfer occurs all the time when human skin comes into physical contact with other objects. Evidence of material transfer can be found in fingerprints or palm prints. A question that has not yet been answered is whether the material transfer is involved in the charge transfer process. In the material transfer of triboelectrification theory, material transfer occurs when part of a long polymer chain is stripped off into small segments that can be transferred. Such a process involves the breaking of at least a covalent bond, and the broken covalent bond will lead to the reaction of the atoms with oxygen, resulting in charges at the material surfaces. In the case of human skin, the material transfer does not involve breaking of a covalent bond. Therefore, it is not clear whether the material transfer could cause charge generation. More efforts need to be made to fully understand the process in the future.

3.2.2. Triboelectric charge transfer in the human body

The electrical structure of the human body is unique because of the biological structure. The outermost part of the human body is the stratum and epidermal, which has a high impedance of up to $10\,M\Omega$ [68]. Beneath the dermal and tissue that have the resistance of approximately $300\,\Omega$ [69]. The high impedance of the skin can act as a dielectric material to generate triboelectric charges by contacting other materials. The difference between the skin and other synthetic insulators is that the skin allows the charges to penetrate at a relatively low breakdown voltage. For skin, the DC breakdown voltage is approximately 500 V, which can be easily achieved by the accumulation of triboelectric or electrostatic charges. Above this breakdown voltage, the charges can penetrate the skin and be conducted through the body. Experimental evidence of such behavior (Fig. 7) has been demonstrated by Zhang and coworkers [70], [71], [72], [73]. Two hypotheses have been developed based on the potentials induced by the triboelectric charges. If the potential is higher than 500 V, then the charges can pass through the skin and be conducted away, resulting in immediately detected electrical signals. Such a hypothesis has been experimentally proven by the measured current of the triboelectric charge flow and the voltage induced on the hand [72] that did not participate in the triboelectrification process. In such a case, the human body acts as a conduit [74]. If the potential is lower than 500 V, then the triboelectric charges will accumulate to build up the potential and then pass through the skin. This hypothesis has been experimentally proven by measuring the discharge of triboelectric charges from the human body. Such a hypothesis also involves the fact that the human body is a network of resistors and capacitors.

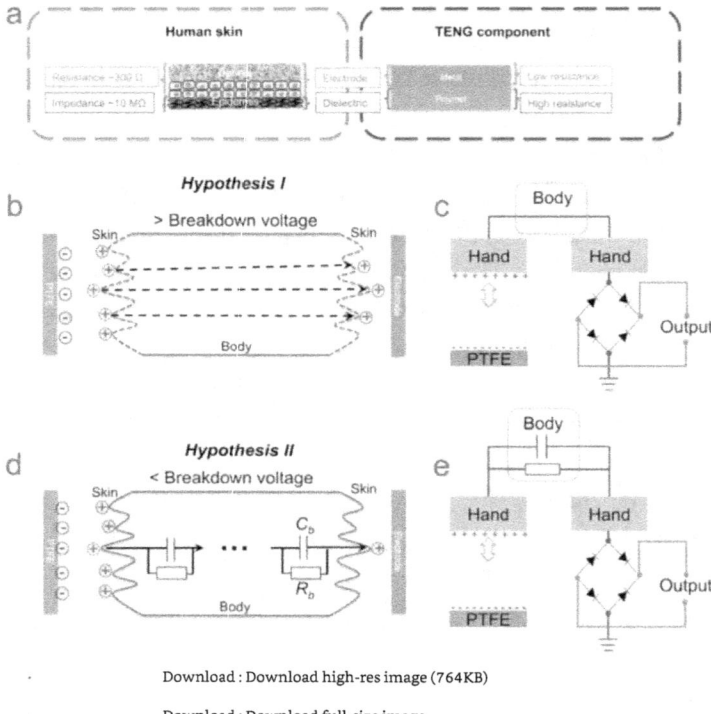

Download : Download high-res image (764KB)

Download : Download full-size image

Fig. 7. Hypothesis of the function of the human body. (a) Human skin and a typical TENG component. The epidermis of the skin and PTFE are dielectric materials with a high impedance or resistance, while the dermis and the electrode deposited on PTFE have low resistances. (b) Hypothesis I, a surface voltage above the skin's breakdown voltage drives charges through the body. (c) Circuit of the H-TENG according to hypothesis I. (d) Hypothesis II, the body acts as capacitors and resistors when the surface voltage is less than the skin breakdown voltage. (e) Circuit of the H-TENG according to hypothesis II.

Ref. [72] Copyright 2018, Elsevier.

Although experimental evidence demonstrating charge transfer through the human body has been shown, there are still some charge transfer behaviors that need to be addressed. For example, if the local potential on the skin is rapidly increased [69], the charges can penetrate easier than in a process with a slower increase. For the triboelectric process, the potential is usually increased in a very short time. Therefore, charges may pass through the skin at a voltage even lower than the breakdown voltage. Such behavior has been found for rapid application of a voltage to the palmar surface, resulting in a capacitive current spike [75].

4. Negative impacts of the triboelectricity of the human body

For a long time before 2012, the impact of the human body's triboelectricity that leads to electrostatic charges was considered negative. Two examples of the negative impact are frizzy hair and the electrostatic discharge of the human body that damages electronics. Products were invented to reduce such an impact of the human body's triboelectricity.

4.1. The hair's triboelectricity

Frizzy hair caused by electrostatic charges is a problem when people try to style their hair. Part of the problem is induced by the shampoo used by consumers. Anionic surfactants, such as ammonium lauryl sulfate, sodium lauroyl sarcosinate, and ammonium laureth sulfate, in shampoos can remove sebum and dirt. However, the use of such shampoos can cause an increase in the negative charge on hair [79]. To reduce the problem, shampoos with cationic, amphoteric, and nonionic surfactants [80] are produced for consumers. New hair care products with specific components have been developed to reduce the surface charge, such as behenamidopropyl dimethylamine [81]. Common components in shampoos and their functions are listed in Table 1 [78].

Table 1. Shampoo formulation components [58], [76], [77], [78].

Shampoo components		Function	Example
Cleaning agent (surfactant)	Anionic	Primary cleansing of the hair with improved removal of lipids	Soap
	Cationic	Provide softness to the hair and improve combability.	Quaternary ammonium salts

		Reduce hair static electricity	
	Amphoteric	Mild cleansers, reduce the tendency of anionics to adsorb onto proteins	Betaines, amphoacetates
	Nonionic	Improve hair manageability. Provide dispersing, emulsifying and detergent properties	Ethoxylated fatty alcohols, tweens, alkyl polyglucosides
Conditioning agent		Impart softness and gloss, reduce flyaway and enhance disentangling facility	–
Special care ingredients		Treat specific hair or scalp conditions, such as dandruff, greasy hair, dermatitis, seborrhea, alopecia, and psoriasis	–
Additives		Contribute to the stability and comfort of	Foam stabilizers, chelating agents,

	the product, adjust the pH and viscosity	viscosity builders (gum, salt, amide)
Preservatives	Reduce possible microbial contamination	–
Aesthetic agents	Provide aesthetics to the shampoo, either color or fragrance	Fragrance, colorants, pearlescent or opacifier agents

Table from ref. [78].

In addition to the components in shampoos, the pH of shampoos also plays a very important role in the charge generation on hair fibers. Alkaline pH has been observed to increase the negative charge density [82] of the hair fiber, which could lead to cuticle damage, especially at pH values higher than 5.5. Cationic ingredients should be added to shampoos if their pH value is above 3.67 [82].

Another way to reduce the electrostatic charge on hair is to apply conditioners. For example, if a shampoo of pH higher than 5.5 has been used, then a low-pH conditioner should be applied to reduce static charge generation [82]. The use of conditioners can reduce friction among hair fibers. Conditioners usually deposit positively charged ions or molecules that possess a natural negative dipole moment [78]. Common components in conditioners and their functions are listed in Table 2 [78].

Table 2. Conditioner formulation components [58], [78].

Conditioner components	Function	Example
Polymers	Increase luster and gloss, reduce the combing force, reduce static electricity (due to hydrophobic agents)	Silicone
Oils/waxes and cationic molecules	Reduce static electricity (due to hydrophobic agents)	Mineral oil, long-chain alcohols, triglycerides, esters
Additives, preservatives, and anesthetic agents	Provide aesthetics to the shampoo, either color or fragrance	Fragrance, colorants, pearlescent or opacifier agents, viscosity builders, pH adjusters, colors

Table from ref. [78].

To style the hair, consumers may use hair dye to change the color. Traditional hair dyes, such as carbon black-based dyes, do not reduce the electrostatic charge on hair. Graphene-based dyes [83] have been recently reported that can significantly reduce the generation of surface charges on hair.

4.2. Electrostatic discharge (ESD)

The daily movements of the human body generate a charge on the body, and the charges could be released suddenly when part of the body comes into contact with (contact discharge) or becomes close to (arc discharge) a conductive object. Such a phenomenon is called electrostatic discharge (ESD). Charge generation on the body was considered to be the result of triboelectrification between, for example, shoes and a resistive floor [84], [85] or cloth on the body and the textiles of chairs [71]. Models such as the capacitance model [86] and multielement model [87] have been developed to simulate the ESD of the human body.

The ESD is electrically characterized by its voltage, current, and waveform. The peak currents are nonlinearly related [88] to the voltage on the human body. The relationship can be described by ESD = 5.144 · ESD 0.6215.

The intensities of the peak currents are usually in the range of several amperes [89].

Since the 1980s, damage of electronic devices due to the human body's ESD [85], [90], [91] has gained much attention. The damage can involve direct failure of physical destruction or indirect failure of false edge sensitivity [90], as well as degradation of semiconductor components [92]. Recently, as **wearable electronics** have come into use, the risk of ESD to electronics has also gained much attention [93]. To protect electronic devices, different types of ESD devices have been invented. The first patented device to reduce the influence of ESD might be the anti-electrostatic garment invented by Gambetti, which was patented in 1965. Such a garment is made as overalls with a cable that conducts charges [94]. However, the garment should be properly designed and used; otherwise, it could also pose an ESD risk to electronics [95]. Currently, wrist straps are mostly used to conduct electrostatic charges away from the human body. In addition to the method of conducting electrostatic charges away, another method that reduces ESD events by introducing a guard electrode [96] between the human body and a target was recently developed. Briefly, a voltage is applied to the guard electrode based on the input from a probe that senses the body's polarity. In this way, the potential difference between the human body and electronics will not be high enough to lead to ESD.

5. Applications that utilize the triboelectricity of the human body

Since the 1660s, scientists have started to discover applications of the human body's triboelectricity (Fig. 8). The milestone of the applications is the invention of the TENG in 2012. Since then, the focus has moved from mainly reducing the influence on electronics of electrostatic charges on the human body to the utilization of triboelectricity for energy conversion and a variety of sensors. Here, we divide the applications into three parts: applications before the 21st century, applications from the beginning of the 21st century to 2012, and applications after 2012.

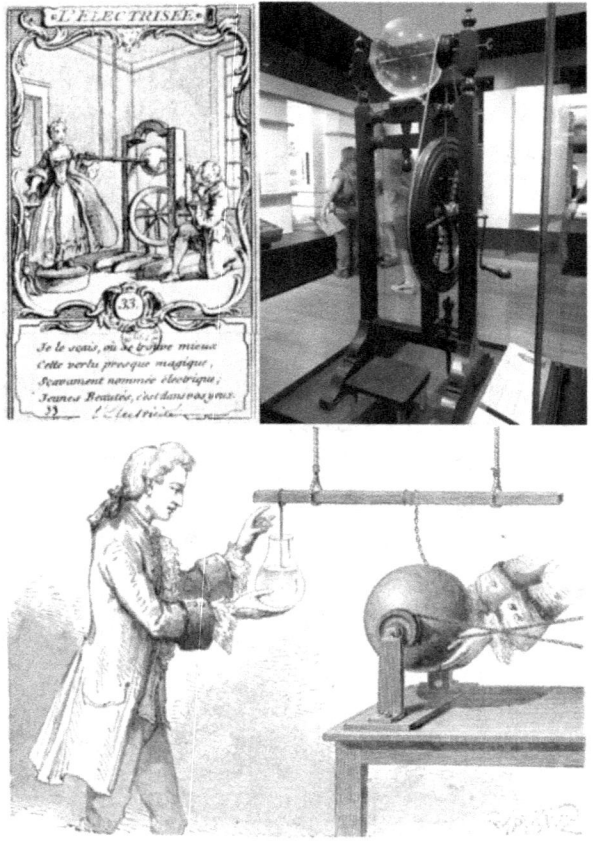

Download : Download high-res image (2MB)
Download : Download full-size image

Fig. 8. Devices used in history to convert the human body's triboelectricity to electricity. Top left: Typical friction machine using a glass globe, common in the 18th century. Top right: Franklin's electrostatic machine on display at the Franklin Institute. Bottom: Discovery of the Leyden jar in Musschenbroek's lab. The static electricity produced by the rotating glass sphere electrostatic generator was conducted by the chain through the suspended bar to the water in the glass held by assistant Andreas Cunaeus. A large charge accumulated in the water, and an opposite charge accumulated in Cunaeus' hand on the glass. When he touched the wire dipped in the water, he received a powerful shock.

All images copied from Wikipedia.com.

5.1. Applications before the 21st century

Before the 21st century, applications that utilized the human body's triboelectricity were mainly focused on **electricity generation**. The mode of electricity generation was similar to the single-electrode mode of TENGs. The first application of the human body's triboelectricity might be the friction machine invented by Otto von Guericke around 1663 [17]. The friction machine generates electricity by rotating and rubbing a sulfur globe with hands. The sulfur globe was later replaced by a glass globe, as suggested by Isaac Newton and made by Francis Hauksbee (Fig. 8). The charges generated by the friction machine can be temporarily stored in a Leyden jar [97] (Fig. 8), invented by Ewald Georg von Kleist, Pieter van Musschenbroek and Andreas Cunaeus. The Leyden jar acts as a capacitor constructed between the human hand and water. In the mid-18th century, Benjamin Franklin (Fig. 8) optimized the friction machine's design and made Franklin's electrostatic machine [98]. Instead of a hand, the machine uses a cloth pad to generate charges that are conducted away with a set of metal needles.

5.2. Applications in the 2000~ 2012 period

Before 2000, especially since the 1980s, the main topic regarding the human body's triboelectricity was the ESD [88], [93], [99], [100], [101], [102], [103]. Studies were conducted to measure the ESD and to develop different methods to reduce its impact. During the period from 2000 to 2012, the focus moved toward sensors that utilized the human body's triboelectricity. An example is the sensor that can sense human walking [104] on different floors such as wooden and ferroconcrete floors [105]. Based on the electrostatic current induced by the human body's triboelectric charges, the human gait can be recognized using remote sensing techniques [106].

Similarly, human hand motions [107] can also

be monitored remotely [108] using the induced electrostatic signals. The common basis of these applications is that they utilize the capacitance between the human body and surrounding environment [109]. No direct touching of the human body with surrounding objects or the ESD process is involved in the applications. A different route that utilizes ESD of the human body can be used to make touch sensors(Fig. 9) [110].

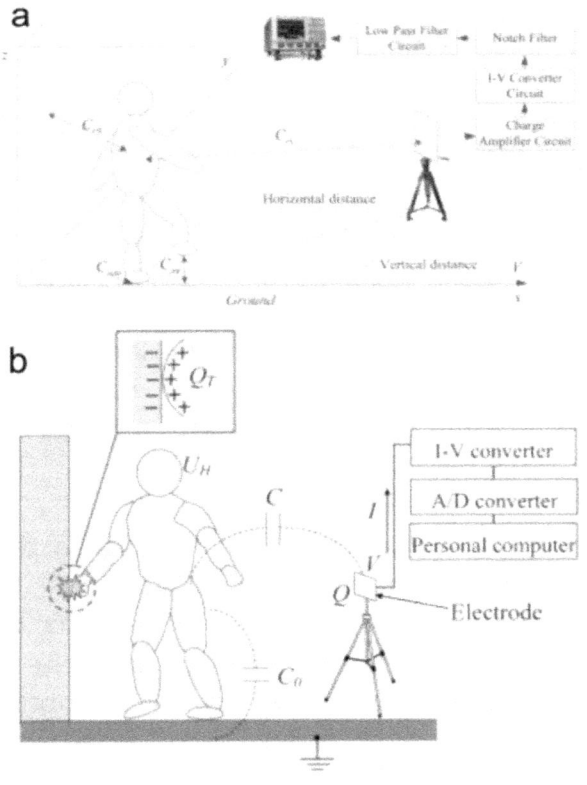

Download : Download high-res image (388KB)

Fig. 9. (a) Schematic diagram of the deployment of a human body motion experiment. Elsevier. (b) Schematic of the measurement system for detecting contact between the subject's hand and the wall.

Part (a) Ref. [106] Copyright 2012. Part (b) Ref. [110] Copyright 2011, Elsevier.

5.3. Applications developed after 2012

In 2012, Wang's group published the first study on TENGs and indicated that such TENGs could be used for harvesting mechanical energy [19]. Such a finding makes triboelectrification a very useful way to produce electrical power. The first TENG that utilizes the human body's triboelectricity was published in 2013 by Yang and coworkers [111]. Recently, many studies have been performed to discover applications of the human body's triboelectricity. Generally, there are two categories of applications: **energy harvesting** and **sensors**.

5.3.1. Utilizing of body motions

From a body motion point of view, the reported TENGs based on the human body's triboelectricity have been applied on different part of the body to harvest energy or sensing. Almost all of the 9 categories of body movements have been adopted in studies including such as flexion and extension, abduction and adduction, etc. Many studies have utilized the hand or fingers movement [112], [113], [114], [115], [116], [117], [118]. However, the operation modes are different, including finger sliding [34], finger pointing [115], finger press [117], [118], [119], and finger bending [114]. Besides, TENGs have also be mounted on the foot [120] or shoes [121] to harvest energy from walking [122], [123], [124].

Because of the diversity of body motions, TENGs could be mounted at different places. Many studies have demonstrated the applications at one or two body parts. However, multiple parts mountable TENGs at different body parts have also been achieved. Lai and co-workers [120] have reported a single-tread-based wearable TENG than can be mounted at the elbow, the knees and on the foot. Xiong and co-workers [125] have reported a textile-based TENGs that can realize durable biomechanical energy harvesting at different places such as shoulders, chest, arm and leg. Wang and co-workers [126] have reported a bioinspired stretchable TENGs that can utilize the muscle's shape change to generate electricity that can be mounted at multiple places of the body. These examples have demonstrated different strategies to take advantages of the results of body motions. Lai's work has utilized the motions at the joints; Xiong's paper have used the body motion induced triboelectrification of the skin; Wang's reports have demonstrated the possibility to use the muscle's contraction and extension.

5.3.2. Energy harvesting

Applications that use the human body's triboelectricity for energy harvesting have a common strategy, that is, to convert body motions into electricity. In most of the studies (Table 3 and the relevant references therein), a single-electrode mode TENG is adopted, and human skin is applied as a triboelectric material to create triboelectrification with a counter dielectric material and generate charges on it. In these cases, the triboelectric charges generated on the human body are ignored. One exception is a report by Zhang and coworkers [72], where they constructed a single-electrode mode H-TENG (Fig. 9) that harvests only the triboelectric charges generated on the human body. The H-TENG produces a higher output power density than other single-electrode TENGs where human skin is involved.

Table 3. TENGs utilizing the human body's triboelectricity.

Empty Cell	Voc (V)	Isc or IscD	P or PD (load)	TENG mode	Ref
PDMS	1000	8 mA/m^2	500 mW/m^2	Single electrode	[111]
PDMS	70	2.7 μA/m^2	–	Single electrode	[131]

PDMS	107.2 (CE) 132 (CF)	0.32 µA (CE) 0.5 µA (CF)	–	Single electrode	[55]
PDMS	70	0.46 µA	135 mW/m$_2$ (800 MΩ)	Single electrode	[132]
PDMS	70	30.2 mA/m$_2$	2.79 W/m$_2$ (150 MΩ)	Single electrode	[133]
PDMS	58.6	–	0.16 W/m$_2$ (40 MΩ)	Single electrode	[134]
PDMS	103	–	4.8 mW/m$_2$ (10 MΩ)	Contact-separation	[127]
PDMS	130	1 µA/cm$_2$	–	Contact-separation	[135]
Silicone	50	6.5 µA/cm$_2$	40 µW/cm$_2$ (1 MΩ)	Single electrode	[136]
Silicone	45	17 µA	19 µW (60 MΩ)	Single electrode	[137]
Silicone	72	18 µA	34.4 µW/cm$_2$ (1 MΩ)	Single electrode	[121]

Silicone	28	0.56 µA	–	Single electrode	[116]
Silicone	200	200 µA	14 mW (1 MΩ)	Contact-separation	[128]
PTFE	320	0.8 µA/cm2	159 µW/m2 (28 MΩ)	Single electrode	[113]
PTFE	570	30 µA	–	Single electrode	[34]
PTFE	600	160 µA	3.3 W/m2 (50 MΩ)	Single electrode	[72]
PTFE	520	120 µA	30 W/m2 (300 MΩ)	Contact-separation	[129]
HBP-fabric	880	1.1 µA/cm2	0.52 mW/cm2 (100 MΩ)	Single electrode	[125]
Kapton (vs. human hair)	103	10.9 µA	60 mW/m2 (1.2 MΩ)	Contact-separation	[56]

The contact-separation mode [127], [128], [129] has also been adopted in some of the studies. Using this mode, the triboelectric charges generated on the human body are included in the energy conversion process. An output power density up to 30 W/m2 could be achieved using this mode (Fig. 10) [129]. However, this mode is less studied due to the relatively more complex construction than the single-electrode mode.

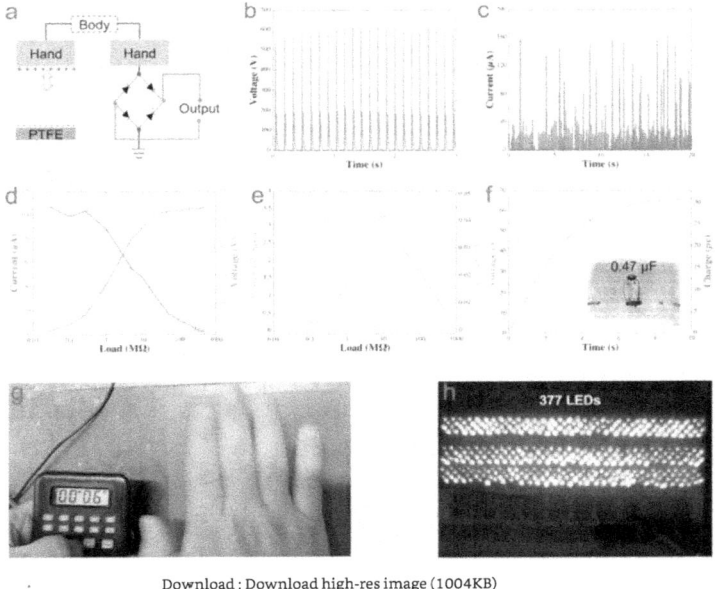

Download : Download high-res image (1004KB)

Download : Download full-size image

Fig. 10. Performance of the H-TENG. (a) Schematic drawing of the circuit of an H-TENG. The function of the body is discussed in the section below. (b) Open-circuit voltage measured for the H-TENG. (c) Short-circuit current measured for the H-TENG. (d) Measured current and voltage vs load resistance. (e) Output power per square meter of the H-TENG vs load resistance. The value was calculated using W=I2R. (f) Charging of a 0.47 μF capacitor by an H-TENG at a hand patting frequency of 6 Hz. The figure shows the voltage and the corresponding charges on the capacitor. (g) Photograph of a timer driven by an H-TENG. h, Photograph of lighting 377 LEDs with an H-TENG.

Ref. [72] Copyright 2018, Elsevier.
From a triboelectric material point of view, polydimethylsiloxane (PDMS), silicone and polytetrafluoroethylene (PTFE) are the most commonly used triboelectric materials in human skin-involved TENGs. Table 3 shows a list of the materials and relevant references. PDMS and silicone are soft materials that can maximize the contact with human skin, resulting in maximized charge transfer. PDMS has an extra advantage that makes it a popular material, the viscoelastic property. Such a property allows PDMS to be easily made in different shapes. PTFE is used mainly because of its significant negative charge affinity, which can lead to high output of the TENGs. In addition to these traditional polymer materials, new materials, such as black phosphorous [125], have also been studied to discover their potential in utilizing to triboelectricity of the human body.

The constructions of the TENGs listed in Table 3 vary depending on the purpose. Generally, they can be classified into two categories: standalone and wearable TENGs. The standalone

TENGs harvest energy, in most cases, from hand motions such as patting. The advantage of such TENGs is the simplicity of the construction. The constructions of wearable TENGs [130] are more complex than those of standalone TENGs. Such TENGs can be attachable to the skin [131] or shoes [121] or wearable as accessories [127] or textiles [125]. For such TENGs, the flexibility and stretchability of the triboelectric materials are crucial. Therefore, textile-like constructions of TENGs with specifically designed fibers have been popularly made [125], [120], [121].

It seems the material choices in literatures that used in TENGs for harvesting energy from human body are not very broad. PDMS, silicone and PTFE are mainly used, despite other materials such as HBP fiber and Kapton have also been used. The reason behind might be the softness of the human skin works better with soft counter materials like PDMS and silicone. In the future, one may expect new types of materials to be used in related studies.

5.3.3. Sensors

TENGs that utilize the human body's triboelectricity are natural body motion sensors because the motions are involved in the triboelectrification process. Different types of sensors, such as force sensors [120], tactile sensors [111], and motion sensors [131], [138], have been developed. The working principles of the single-electrode and contact-separation modes of the TENGs that utilize the human body's triboelectricity make them respond differently to different applied forces. However, the electrical signals are not linearly related to the applied force [120], requiring further studies to optimize either the material selection or the construction of the TENG. A tactile sensor also senses the force that is applied to a TENG; however, the sensitivity should be much higher than that of a force sensor. Yang and coworkers [111] developed a tactile sensor with a sensitivity of 0.29 V/kPa (Fig. 11). An 8 × 8 matrix tactile sensor with a pixel size of 3 mm × 3 mm can sense the force difference at different locations. Kim's group fabricated a system [117] combining single-electrode-mode TENGs (S-TENGs) and field-effect transistors (FETs) to sense finger touching (Fig. 12). Such a tribotronic device has a sensitivity of ≈2% kPa^{-1} and a detection limit of <1 kPa. Another dual-mode sensor [115] was recently developed for sensing finger touching in both a high pressure range (10–120 kPa) and a low pressure range (<10 kPa) with a sensitivity of 1.04 V/kPa.

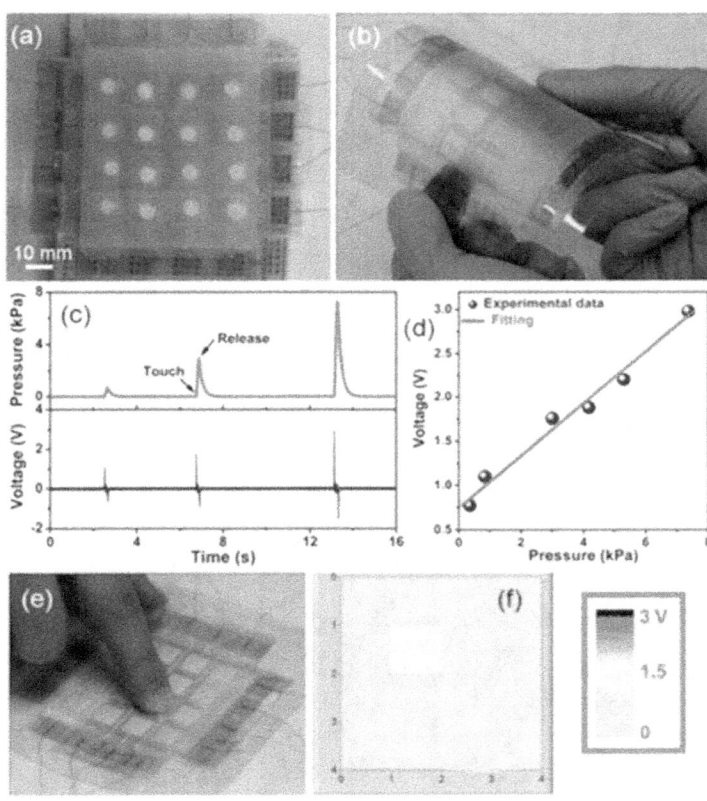

Fig. 11. (a) Photograph of a device on a red LED array, exhibiting the transparent feature. (b) Photograph of the bent device. (c) Output voltages of one device in the matrix under different pressures. (d) Output voltage as a function of applied pressure. The red line corresponds to the linear fitting function. (e) Photograph of the matrix touched by a human finger. (f) Measured positive output voltage map for the touching in e.

Ref. [111] Copyright 2013, American Chemical Society.

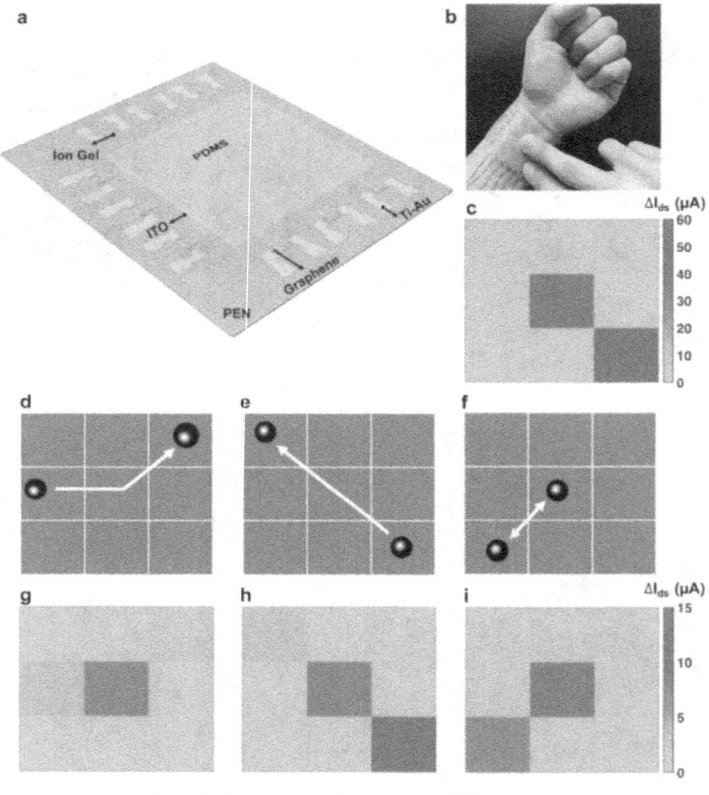

Fig. 12. Graphene tribotronic array. a) Schematic depiction of a 3 × 3 graphene tribotronic array. b, c) Photograph of the tribotronic array mounted on the wrist with a double finger touch (b), and corresponding spatial map in terms of current modulation, ΔIds (c). (d–f) Schematic depiction of the movement of a ball over the PDMS friction layer, and (g–i) corresponding spatial maps in terms of ΔIds.

Ref. [117] Copyright 2016, John Wiley and Sons, Ltd.

More recently, a new type of 3D touch pad [139] has been developed to enhance the human-machine interaction. Such a 3D structure constructed by a multi-channel positioning layer and a single-channel pressure sensing layer. The working principle of the 3D touch pad is shown in Fig. 13. In brief, the touch of positive charged finger will drive the positive charges bounded on the electrode and shielding layer to the ground to keep a neutralization. Further pressing of the finger will lead to the triboelectrification at the sensing layer. There two procedure produces different signals for processing. The results in the paper will lead to a universal tactile sensing strategy for robotics and human-machine interactions.

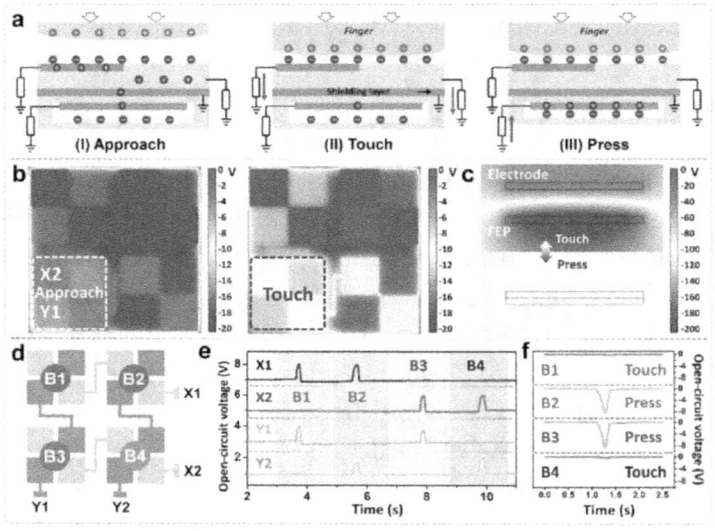

Fig. 13. Working principles of the 3D-TTP. (a) Cross-sectional view of a sensing unit and charge distribution within an Approach-Touch-Press process. (b) Simulation by COMSOL to elucidate the potential variation of the positioning layer and its influence on the other electrode. (c) Potential simulation by COMSOL of a pressure sensing unit shielded from the upper positioning part. (d) and (e) Positioning principle based on the synchronous signals from the intersected rows and columns. (f) Signals of the pressure sensing part synchronous to the upper positioning signals.

Ref. [139] Copyright 2020, Elsevier.

Another type of sensor is dedicated to sensing body motions that is different from sensing forces because it has the requirement of recognizing or identifying different motions [30]. A simple approach to recognizing different motions is to put multiple sensors on different body parts [140], [141], and the response of one of the sensors indicates the motion of the body part. Such an approach is effective but requires a complex connection of sensors and actuators. Zhang and coworkers [71] reported a new way of sensing body motions by reading the triboelectric signals generated at the moving body part (**Fig. 14**). In this way, one has the possibility of using a simple circuit to identify different body motions. However, due to the complexity of body motions, there is a need for assistance from computer science, such as machine learning, to improve the performance of the sensor.

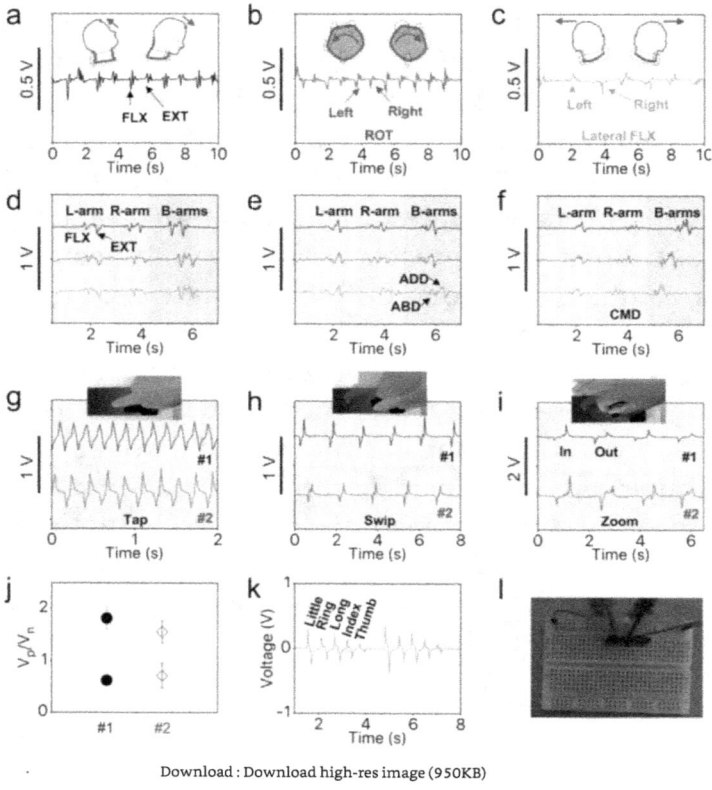

Fig. 14. Sensing of head, arm and finger motions. (a) Voltage changes during flexion/extension. (b) Voltage changes during rotation. (c) Voltage changes during lateral flexion. The drawings show the movements of the head. The dashed circles show the locations of contact between the skin and collar. (d) Arm flexion/extension. (e) Arm adduction/abduction. (f) Arm circumduction. d) Waist flexion/extension. The results of three replicates are shown. (g) Voltage

of finger tapping on mobile phones. h) Voltage of finger swiping on mobile phones. (i) Voltage of zooming in and out finger movements on mobile phones. (j) Ratio between positive (Vp) and negative (Vn) potentials. (k) Voltage of finger tapping on a polytetrafluoroethylene (PTFE) film. (l) LED lit by finger tapping on a mobile phone screen. The positions of replicates are shifted up or down along the y axis to show the results clearly.

Ref. [71] Copyright 2019, Elsevier.
In addition to tactile and motion sensors, there are other types of sensors with more specific aims utilizing the human body's triboelectricity. One example is a TENG attached to fingers to sense finger joint motion and predict the bending angles. A sensitivity of approximately 0.3 pF/° has been achieved. Another example is a control disk [34] that sends different codes to realize **smart home** control and password **authentication access** control.

6. Perspectives

TENGs have attracted much attention from researchers in different areas due to their broad range of applications. The study of the human body's triboelectricity is however less studied compared to other areas, such as **energy harvesting** from different resources, although increasing interest has been observed recently in the literature.

In addition to the above reviewed studies, there are many other studies that one can perform in the future.

- 1)
    Mechanism of the skin's triboelectrification. The epidermal layer of human skin has many chemical molecules that contribute to its triboelectric properties. How these molecules act under different conditions is not known. In addition, how skin conditions, such as wrinkles and elasticity, influence the triboelectricity is also unclear.

2)
Biomedical applications utilizing the human body's triboelectricity. A person moves differently when he or she is experiencing different diseases, which can generate different triboelectric signals. By analyzing the signals, a doctor can obtain a better diagnosis. An example is to sense the body's triboelectric signals for a person with Parkinson's disease [71]. By analyzing the information of the signals, e.g., frequency and intensity, a doctor can predict the progress of the disease and give proper treatment.

3)
Human-machine interactions utilizing the human body's triboelectricity. There are studies to create human-machine interactions

by triggering TENGs with body motions. However, using the human body's triboelectricity would be more precise and may communicate more complex information.

4)

Aging of the human skin. There is no knowledge about how human skin aging relates to its triboelectrification. However, as the skin is experiencing triboelectrification processes all the time, there could be a potential relationship between the aging and triboelectrification.

The study of the human body's triboelectricity is not a simple task, requiring multidisciplinary collaborations with scientists from physics, electronics, chemistry and biology. To utilize the human body's triboelectricity, one may need input from computer science.

7. Conclusions

Ever since the discovery of triboelectrification, the human body's role has been involved. Many scientists have made many efforts to study and utilize the human body's triboelectricity. We have reviewed here the studies that focus on the human body's triboelectricity from the beginning of the discovery. A brief history of triboelectrification has been given along with information about important people related to the human body's triboelectricity. Mechanisms of triboelectrification have also been briefly reviewed, showing the differences in the theories.

The core of this paper is to review the studies that are directly related to the human body's triboelectricity. Studies at different times and from different areas, such as cosmetics, electronics protection, energy harvesting, and sensors, have been reviewed. Perspectives for future studies have also been given, focusing on the mechanisms of human skin triboelectrification and the applications that can be achieved by multidisciplinary collaborations.

The paper gives a systematic overview of the human body's triboelectrification and applications that are directly related. The information provided here could promote the study of the human body's triboelectrification, which may have great importance in many applications, such as biomedical devices and artificial intelligence.

HIGH SCHOOL

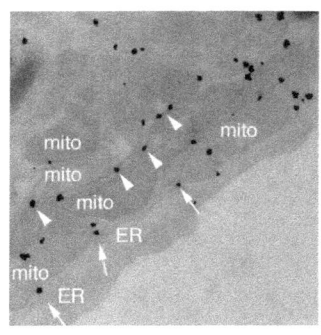 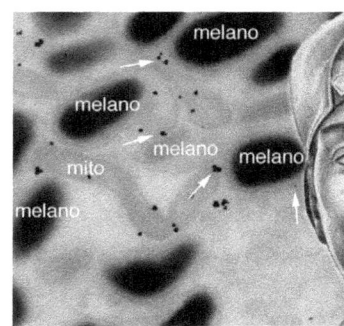

Thoughts on Melanosomes and Melanin being attached to "ELECTRON TRANSPORT CHAIN UNITS"... HOW ABOUT THE FACT THAT CAPACITORS WERE USED LIKE MAGIC WANDS IN ANCIENT EGYPT???

KEEP IN MIND....

THIS IS JUST FOR YOUR MIND! THINK AWAY!

ELECTRICAL CAPACITOR OR MAGICK WANDS TOMATO/TOMATOE

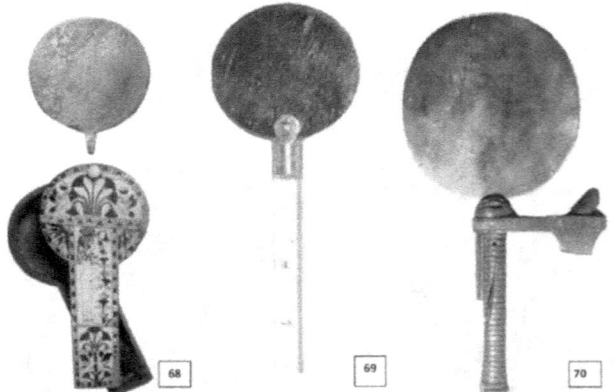

68 METAL DISKS HAD SPECIAL CASES

69 ELECTRICAL CAPACITOR PURCHASED AT EDMUNDS SCIENTIFIC IS DESIGNED TO CAPTURE STATIC ELECTRICITY TO BE DISCHARGED SOMEWHERE ELSE

70 WOODEN HANDLE MAKES METAL DISK AN ELECTRICAL CAPACITOR

CHASEDUQUESNAY

JUST IN CASE YOU NEED TO SEE THE WANDS

CHASEDUQUESNAY

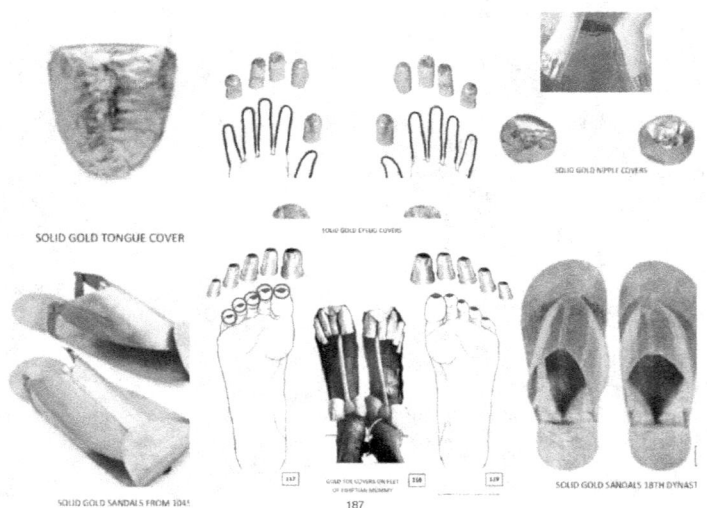

THE MELANIN BOOK

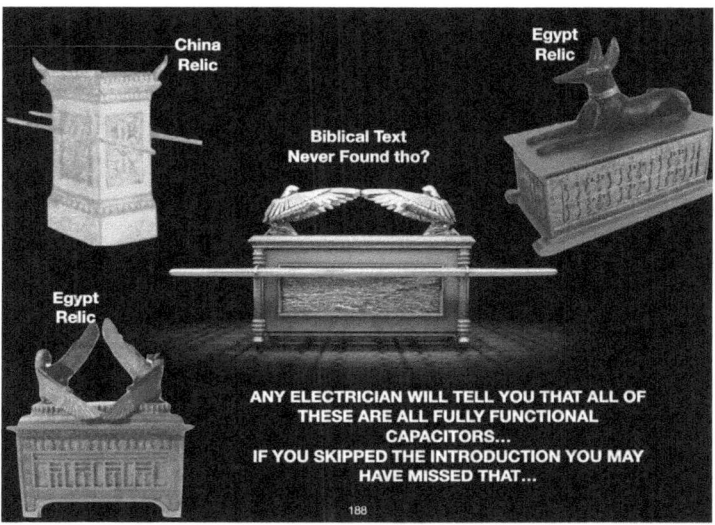

There are Amber medallions going back 12,000 years and some how there seems to be a 'Blackout' on Electricity usage by the people who built pyramids....???

I believe it is the Horus and Set book, in which we discuss the second Sun, or the Electric Field projected here, by the Sun through our Plasma Bubble Lens. I know we discuss it in the TIME TABLE but... people may not have linked the two... The 60 mile high electric field that we call Day Time is a manifestation of the Sunlight and our Plasma Lens (Magnetosphere).

There are many theories on Melanin, however none of them will be completely correct, until we factor in the Electric Field. Our current understanding of Light itself must shift, when we get this idea down pact about the electric field we live in.

Mind you I have already told you the keys in parts... Put it together...
It's simple aint it, but quite clever...
Most of you been trying to write rhymes for years,
But weak ideas irritate my ears.... - Rakim

It is my theory that we won't even fully understand Masonry, until we understand Egypt's Electromagnetic Mastery. In my heart of hearts I truly believe this "the Craft"...
Electromagnetism the Quest to understand God.

DOLLARD BUILDS...

Φ	TOTAL MAGNIFICATION WEBER	Ψ	TOTAL DIELECTRIFICATION COULOMB
Q	TOTAL ELECTRIFICATION PLANK		A = AREA/SPACE T = TIME
Φ/A	DENSITY OF MAGNETIC INDUCTION PER CM^2		
Ψ/A	DENSITY OF ELECTRICAL POTENTIAL PER CM^2		

φ/A^2 TOTAL DENSITY OF ELECTRICITY PER CM ^{2}

φ/T WORK / JOULE = W

Ψ/T MAGNETO MOTIVE FORCE / AMPERE = I

Φ/T ELECTROMOTIVE FORCE / VOLT = E

φ/T^2 POWER OR ACTIVITY / WATT = P

VOLTS SHOW MAGNETIC CONSUMPTION AND THE MAGNETO MOTIVE FORCE REPRESENTS DIELECTRIC CONSUMPTION

$\Phi/I = L$ MAGNETIC INDUCTANCE HENRY

$\Psi/E = C$ DIELECTRIC CAPACITY FARADAY

E/I = Z IMPEDANCE OHM   I/E = Y ADMITTANCE SIEMENS

L/T = R RESISTANCE PER SECOND-HENRY-DESTRUCTION OF ELECTRICITY OHM

C/T = G CONDUCTANCE - CREATION OF ELECTRICITY - SIEMENS - MOHS

$L-C=T \longrightarrow \sqrt{LC}-T-F°$ **HERTZ**

DUALITY

DIMENSION OF TIME	DIMENSION OF SPACE
T = SECONDS	$L = CM^2$
FORWARD IN TIME +1T	L^{+1} OUTERSPACE
BACKWARDS IN TIME -1T	L^{-1} INNERSPACE
ADDITIVE	MULTIPLICATIVE

$X^2 = -1$
$X^2 = +1$

STEINMETZ UNIFIED THE 4 DIMENSIONS OF SPACE AND TIME BY CONVERTING TIME AND SPACE TO LIGHT SECONDS AND USING THE GOLDEN RATION FOR DISTRIBUTION

MAGNETIC FORCE LINES ARE 360 DEGREE PERFECT CIRCLES AND DIELECTRIC FORCE LINES RUN CONDUCTOR TO CONDUCTOR, POLE TO POLE...

IN A MAGNETIC FIELD THE ELECTRONS AT THE END OF THE DIELECTRIC LINE OF FORCE WILL SNAKE AND EAT HIS OWN TAIL BECOMING A CIRCLE

THE MELANIN BOOK

ELECTROMAGNETISM IS NOT BOUND TO TIME/ SPACE BECAUSE THEY AREN'T REAL THINGS

MAGNETISM WORKS IN OUR DIMENSION, OUTER SPACE IF YOU WILL AND DIELECTRIC FORCE BEING MICROSCOPIC AT BEST WORK ON INNER SPACE!

 THE ELECTRICITY IS TRAVELING IN THE WIRE AND THE MAGNETIC FORCE IS ON THE OUTSIDE OF THE WIRE

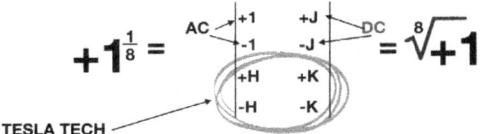

TESLA TECH

WAKEY WAKEY YOU SEE THE 4 SETS OF TWINS? PLUS THE 1 & 8 SIGNATURE

COLLEGE

Plasma based Crystal disc, fitted with integrated circuits as well as gates and channels fitted in a Capacitor (membrane). The full body capacitors are also fitted with a wide variety of antenna and transformers. These Nanobots have 2000+ alternators and motors in them for internal power. Each of these cells are Electrochemical with thousands of smaller nano electric machines.

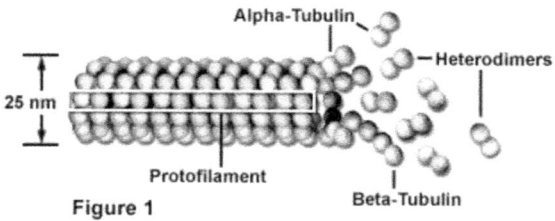

Figure 1

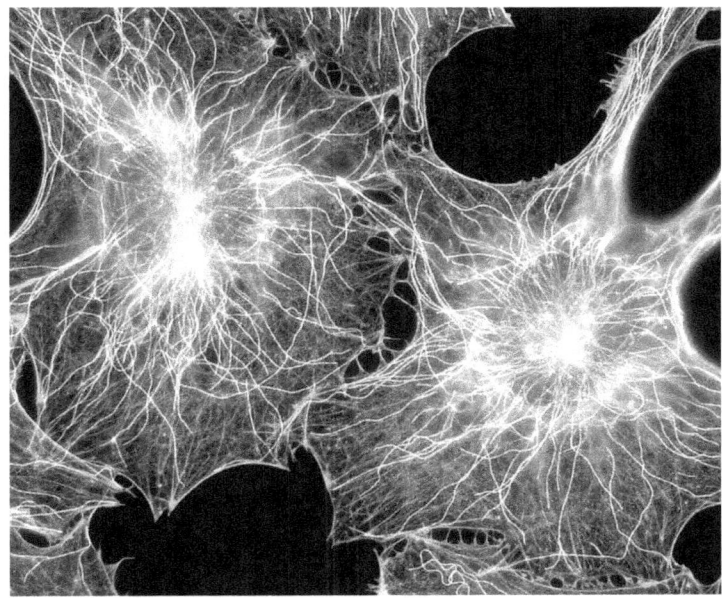

Investigation of the Electrical Properties of Microtubule Ensembles under Cell-Like Conditions

Aarat P. Kalra,1 Sahil D. Patel,2 Asadullah F. Bhuiyan,2 Jordane Preto,1 Kyle G. Scheuer,2 Usman Mohammed,3 John D. Lewis,4 Vahid Rezania,3 Karthik Shankar,2,* and Jack A. Tuszynski1,4

Abstract

You can read the **Full Article Later** these are just excerpts from this article, and Wiki

Cells can go from a few to a few thousand Microtubules. We observed a decrease in electrical resistance of solution, with charge transport peaking between 20–60 Hz in the presence of microtubules, consistent with recent findings that microtubules exhibit electric oscillations at such low frequencies. We were able to quantify the capacitance and resistance of the microtubules (MT) network at physiological tubulin concentrations to be 1.27×10^{-5} F and 9.74×10^4 Ω. Our results show that in addition to macromolecular transport, microtubules also act as charge storage devices through counterionic condensation across a broad frequency spectrum. We conclude with a hypothesis of an electrically tunable cytoskeleton where the dielectric properties of tubulin are polymerisation-state dependent. They have been modelled as nanowires capable of enhancing ionic transport, and simulated to receive and attenuate electrical oscillations. In solution, MTs have been shown to align with applied electric fields. Recently, MTs have also been modelled as the primary cellular targets for low-intensity (1–2 V), intermediate-frequency (100–300 kHz) electric fields termed TTFields (tumour-treating electric fields) that inhibit cancer cell proliferation, in particular glioma. Indeed, MTs have been reported to decrease buffer solution resistance, leading to a conductance peak at frequencies close to the TTField regime.

Our work, which utilizes cell-like tubulin

and ionic concentrations for the first time, indicates a cellular role for microtubules as wires that store charge. Neuronal environments where MTs are spontaneously nucleated from free tubulin, such as growth cones, experience large capacitance changes over short bursts of time. This ability significantly impacts the action potentials that are known to depend strongly on the local charge distributions [56]. Additionally, ionic movement across the MT wall enhances their roles as attenuators of local cation distributions. In nonneuronal environments, transient ionic currents around a MT during mitosis could impact MT dynamics and potentially influence the chromosome segregation. Specifically, Ca^{2+} ion storage/flow about an MT triggers its depolymerisation, whereas waves of Mg^{2+} or lowering in the local pH (increasing H^+) leads to MT stabilisation [57,58]. The attraction of Zn^{2+} or Mn^{2+} ions in the vicinity leads to the formation of two-dimensional tubulin polymers [59,60]. Properties of the cytoplasm such as polarisability and relative permittivity get severely attenuated because of the presence of MTs in the vicinity. Because of the polymerisation state of tubulin-altering solution capacitance, our findings implicate a temporal evolution of capacitance and ionic flows as the ratio of MTs to free unpolymerised tubulin changes [61,62,63]. MT lattice defects, which occur when a tubulin dimer is missing in an MT wall [64,65], cause a large ionic flux to develop at the defect site. Spatiotemporal charge distribution shifts are also critical at the MT end, where fluxes form because of sudden changes due to the polymerisation/depolymerisation of the MT. Free/polymerised tubulin hence regulates local and global electrical properties, creating spatially dynamic gradients of charge storage and flux. We envision a cytoskeleton that, in addition to transporting macromolecules, stores and transports ionic signals and electrical information across the cytoplasm (Figure 9a,b).

Our findings can be coupled with a vast array of bio-nanodevices that utilises MTs and MAPs (microtubule-associated proteins) for construction of bio-nanotransporters and bio-actuators [66,67,68,69,70]. Under specific conditions, MAP-MT systems are capable of repositioning macromolecules [71,72], directionally transporting microtubules [15,73] and even drive their movement within zero-mode waveguides [74] and inorganic nanotubes [75]. Storage of electrical charge and its transport along MTs can be coupled to such cutting-edge mechanical MAP-based devices to develop a wide range of nano-actuators and nano-sensors.

When compared to cells, the rates of MT nucleation and polymerisation are significantly lower in BRB80. This difference can be attributed to the absence of MAPs and macromolecular crowding [76,77]. Mammalian cells contain high concentrations of K+ ions (140–300 mM) [27,28], which, in addition to MAPs and molecular crowding agents, will be included in a future study to attain physiological equivalence. We also note that the effect of PTMs (post-translational modifications) on the electrical properties of microtubules has not yet been explored. PTMs involve the addition of residues such as phosphate and glutamate that locally influence counterionic condensation around the outer microtubule surface.

We are in the process of performing DC (direct-current) measurements, determine the contribution of MTs to impedance relaxation time and evaluate the voltage dependence of capacitance on MT-containing solutions. Interestingly, this aspect has been discussed previously: the inductance of a single protofilament is calculated to be < 1 fH.

We used EIS to compare the complex impedance of MT- and tubulin-containing solutions. A physiologically relevant, high ionic strength buffer (BRB80) created a high noise, low impedance background, which was countered through the use of physiological concentrations of tubulin. While the presence of MTs increased solution capacitance, unpolymerised tubulin did not have any appreciable effect. In a study that is the first of its kind to the best of our knowledge, we determined the capacitance and resistance of the MT network at physiological tubulin concentrations to be 1.27×10^{-5} F and 9.74×10^4 Ω. These values correspond to an effective resistance per unit volume of 3.71×10^{10} Ω/L and effective capacitance per unit volume of 7.65 F/L. We envision a dual electrical role for MTs in the cell, that of charge storage devices across a broad frequency spectrum (acting as storage locations for ions), and of charge transporters (bionanowires) in the frequency region between 20 and 60 Hz. Our findings also indicate that the electrical properties of tubulin dimers change as they polymerise, revealing the potential impact of MT nucleation and polymerisation on the cellular charge distribution. Our work shows that by storing charge and attenuating local ion distributions, microtubules play a crucial role in governing the bioelectric properties of the cell. - NLM

Microtubules can be as long as 50 micrometres, as wide as 23 to 27 nm[2] and have an inner diameter between 11 and 15 nm. [3] They are formed by the polymerization of a dimer of two globular proteins, alpha and beta tubulin into protofilaments that can then associate laterally to form a hollow tube, the microtubule.[4] The most common form of a microtubule consists of 13 protofilaments in the tubular arrangement.

Microtubules are one of the cytoskeletal filament systems in eukaryotic cells. The microtubule cytoskeleton is involved in the transport of material within cells, carried out by motor proteins that move on the surface of the microtubule.

Microtubules play an important role in a number of cellular processes. They are involved in maintaining the structure of the cell and, together with microfilaments and intermediate filaments, they form the cytoskeleton.

Microfilaments, also called actin filaments, are protein filaments in the cytoplasm of eukaryotic cells that form part of the cytoskeleton. They are primarily composed of polymers of actin, but are modified by and interact with numerous other proteins in the cell. Microfilaments are usually about 7 nm in diameter and made up of two strands of actin. Microfilament functions include cytokinesis, amoeboid movement, cell motility, changes in cell shape, endocytosis and exocytosis, cell contractility, and mechanical stability.

Microfilaments are flexible and relatively strong, resisting buckling by multi-piconewton compressive forces and filament fracture by nanonewton tensile forces. In inducing cell motility, one end of the actin filament elongates while the other end contracts, presumably by myosin II molecular motors.

Intermediate filaments (IFs) are cytoskeletal structural components found in the cells of vertebrates, and many invertebrates.[1][2][3] Homologues of the IF protein have been noted in an invertebrate, the cephalochordate Branchiostoma.[4]

Intermediate filaments are composed of a family of related proteins sharing common structural and sequence features. Initially designated 'intermediate' because their average diameter (10 nm) is between those of narrower microfilaments (actin) and wider myosin filaments found in muscle cells, the diameter of intermediate filaments is now commonly compared to actin microfilaments (7 nm) and microtubules (25 nm).[1][5] Animal intermediate filaments are subcategorized into six types based on similarities in amino acid sequence and protein structure.[6] Most types are cytoplasmic, but one type, Type V is a nuclear lamin. Unlike microtubules, IF distribution in cells show no good correlation with the distribution of either mitochondria or endoplasmic reticulum. - wiki

The alpha and beta tubulin are built on aromatic amino acids and phosphates... Sound familiar? The repetitive arrangement of tubulin dimers is like the chess board, like Melanin remember? Pages 311-314 in L'Goat book...? These microtubules are complex filaments!!! They work very similar to Melanin just in reverse. Melanin blocks light absorption, microtubules channel and amplify light with the same crystal lattice structure of repeating units. The Phosphates and Aromatic Amino Acids amplify the initial signal from the cell body down the axon, there is paperwork on that. The question for you is what are the filaments do in ALL THE REST OF YOUR CELLS?

<u>Are we ready to just say damn it your right EnQi!!! You have found the SOUL. Self Organizing Universal Light is the basis and medium of consciousness, facilitated by Phosphorus...</u>

Filaments are for light! Period! Did you know that ATP has a twin? GTP... GTP does the same thing ATP does but is used for specific jobs like Microtubule activity!

GTP, in combination with ribulose 5-phosphate, are the precursor compounds for the synthesis of riboflavin (vitamin B2).

It is essential to the formation of two major coenzymes, flavin mononucleotide and flavin adenine dinucleotide. These coenzymes are involved in energy metabolism, cellular respiration, and antibody production, as well as normal growth and development. The coenzymes are also required for the metabolism of niacin, vitamin B6, and folate. Riboflavin is prescribed to treat corneal thinning, and taken orally, may reduce the incidence of migraine headaches in adults.

Guanosine-5'-triphosphate (GTP) is a purine nucleoside triphosphate. It is one of the building blocks needed for the synthesis of RNA during the transcription process. Its structure is similar to that of the guanosine nucleoside, the only difference being that nucleotides like GTP have phosphates on their ribose sugar. GTP has the guanine nucleobase attached to the 1' carbon of the ribose and it has the triphosphate moiety attached to ribose's 5' carbon.

It also has the role of a source of energy or an activator of substrates in metabolic reactions, like that of ATP, but more specific. It is used as a source of energy for protein synthesis and gluconeogenesis.

GTP is essential to signal transduction, in particular with G-proteins, in second-messenger mechanisms where it is converted to guanosine diphosphate (GDP) through the action of GTPases.

GTP is involved in energy transfer within the cell. For instance, a GTP molecule is generated by one of the enzymes in the citric acid cycle. This is tantamount to the generation of one molecule of ATP, since GTP is readily converted to ATP with nucleoside-diphosphate kinase (NDK).

During microtubule polymerization, each heterodimer formed by an alpha and a beta tubulin molecule carries two GTP molecules, and the GTP is hydrolyzed to GDP when the tubulin dimers are added to the plus end of the growing microtubule. Such GTP hydrolysis is not mandatory for microtubule formation, but it appears that only GDP-bound tubulin molecules are able to depolymerize. Thus, a GTP-bound tubulin serves as a cap at the tip of microtubule to protect from depolymerization; and, once the GTP is hydrolyzed, the microtubule begins to depolymerize and shrink rapidly.

The translocation of proteins into the mitochondrial matrix involves the interactions of both GTP and ATP. The importing of these proteins plays an important role in several pathways regulated within the mitochondria organelle, [4] such as converting oxaloacetate to phosphoenolpyruvate (PEP) in gluconeogenesis. - wiki

Plasma based Crystal disc, fitted with integrated circuits as well as gates and channels fitted in a Capacitor (membrane). The full body capacitors are also fitted with a wide variety of antenna and transformers. These Nanobots have 2000+ alternators and motors in them for internal power. Each of these cells are Electrochemical with thousands of smaller nano electric machines.

We now have to add the filaments into this Crystal Disc we call a Body Cell or Somatic Cell. It make sense and it's easy to understand what they are doing in Neurons, the amplify the electromagnetic signal generated in the cell body, until it can be translated into chemicals. Got it! In the dendrite they perform the task just backwards, they allow the incoming chemicals to be translated back into electromagnetic signals for the cell body. Digital/Analogue Binary Coding...

The question you have to ask yourself is WHAT ARE THIS LIGHT HARVESTING AND AMPLIFYING FILAMENTS DOING IN ALL OF OUR CELLS?

ATP & GTP are both energy sources, phosphate oxidation is the reaction, and we know what that's about right? Sooooooooo
Let's go to the Bible real quick...

God is Light.

The kingdom of God is within us.

The kingdom of Light is within us?

We know through painstaking study, that Light is information, information is knowledge etc...

In the Human Body it seems light is it's power and information source!

The ability to do work and the description of the work to be done.

This would mean God = Light, Light = Knowledge so God = Knowledge.

If P then Q.

MICROTUBULE DATA

Investigation of the Electrical Properties of Microtubule Ensembles under Cell-Like Conditions

Aarat P. Kalra,[1] Sahil D. Patel,[2] Asadullah F. Bhuiyan,[2] Jordane Preto,[1] Kyle G. Scheuer,[2] Usman Mohammed,[3] John D. Lewis,[4] Vahid Rezania,[3] Karthik Shankar,[2,*] and Jack A. Tuszynski[1,4]

Abstract

Microtubules are hollow cylindrical polymers composed of the highly negatively-charged (~23e), high di- pole moment (1750 D) protein α, β- tubulin. While the roles of microtubules in chromosomal segregation, macromolecular transport, and cell migration are relatively well-understood, studies on the electrical prop- erties of microtubules have only recently gained strong interest. Here, we show that while microtubules at physiological concentrations increase solution capacitance, free tubulin has no appreciable effect. Further, we observed a decrease in electrical resistance of solution, with charge transport peaking between 20–60 Hz in the presence of microtubules, consistent with recent findings that microtubules exhibit electric oscil- lations at such low frequencies. We were able to quantify the capacitance and resistance of the micro- tubules (MT) network at physiological tubulin concentrations to be 1.27×10^{-5} F and 9.74×10^{4} Ω. Our results show that in addition to macromolecular transport, microtubules also act as charge storage devices through counterionic condensation across a broad frequency spectrum. We conclude with a hypothesis of an electrically tunable cytoskeleton where the dielectric properties of tubulin are polymerisation-state dependent.

Keywords: microtubules, bioelectricity, bionanowires, neuronal charge storage, impedance spectroscopy, cytoskeleton

1. Introduction

Microtubules (MTs) are cylindrical polymers composed of the heterodimers of protein α, β- tubulin that play a variety of well-recognised intracellular roles, such as maintaining the shape and rigidity of the cell, aiding in positioning and stabilisation of the mitotic spindle for allowing chromosomal segregation, acting as 'rails 'for macromolecular transport and forming cilia and flagella for cell movement. Since the tubulin dimer possesses a high negative electric charge of ~23e and a large intrinsic high dipole moment of ap- proximately 1750 D [1,2], MTs have been implicated in electrically-mediated biological roles [3,4,5,6]. They have been modelled as nanowires capable of enhancing ionic transport [7,8], and simulated to receive and attenuate electrical oscillations [4,9,10,11]. In solution, MTs have been shown to align with applied electric fields [2,12,13,14,15,16]. Recently, MTs have also been modelled as the primary cellular targets for low-intensity (1–2 V), intermediate-frequency (100–300 kHz) electric fields termed TTFields (tumour- treating electric fields) that inhibit cancer cell proliferation, in particular glioma [17,18,19]. Indeed,

MTs have been reported to decrease buffer solution resistance [12,13], leading to a conductance peak at frequen- cies close to the TTField regime [20]. While these studies show that MTs are highly sensitive to external electric fields, answers to the questions 'How do MTs effect a solution's capacitance? 'and 'What is the ca- pacitance of a single MT? 'are still elusive and crucial to the determination of the dielectric properties of living cells. The tubulin concentration in mammalian cells varies in the micromolar range (~10–25 µM) [21,22]. In vitro, polymerizing tubulin at such high concentrations can lead to the formation of entangled networks, confounding quantification of the individual MT response to electric fields. Electro-rotation, dielectrophoresis and impedance spectroscopy are thus performed using low concentrations of tubulin, in the nanomolar regime, to enable robust observation of individual MTs.

MT formation and stability are known to be optimal in buffers with ionic strength between 80 and 100 mM [23,24]. A background of BRB80 (which consists of 80 mM PIPES, 2 mM $MgCl_2$ and 0.5 mM EGTA, containing ~46 mM $PIPES^{2-}$, ~36 mM $PIPES^-$, ~68 mM Cl^-, ~160 mM K^+, and ~2 mM Mg^{2+} [2]), is thus used to study the dynamics and mechanical properties of MTs. To study their electrical properties however, the usage of such high ionic-strength solutions has historically been problematic because any di- electric attenuation caused by MTs is overwhelmed by the noise and high conductivity from the back- ground. In the low-frequency regime (1–100 kHz), two approaches have thus far been used to estimate the dielectric properties of MTs and tubulin. One is to electrically observe low concentrations of MTs (tubulin concentration in the nanomolar regime) in the presence of low ionic strengths [12,13,20,25,26]. Such stud- ies overlook the intrinsic ionic concentration of mammalian cytosol, which varies between 200 to 500 mM depending on the cell type [27,28]. Another approach to electrically interrogate MTs is to dry them: the conductivity of the buffer is nullified by evaporation, leaving polymeric tubulin behind [29,30]. In a physi- ological situation however, MTs are solvated by the highly conductive and noisy cytosol.

Here, we report on our efforts overcome the barrier posed by a high ionic strength by performing electro-chemical impedance spectroscopy (EIS) on cellular concentrations of tubulin. We have been able to suc- cessfully observe differences in impedance using a background of BRB80 itself. Surprisingly, we find that MTs increase the solution capacitance of BRB80 whereas free tubulin does not, implicating a difference in electrical properties based only on the morphology of this protein solute. We also report a 'reversal 'in the resistive behaviour of MTs compared to BRB80, with a reduction in solution resistance peaking in the 20– 60 Hz region, a finding consistent with recent reports showing that polymerised tubulin quasi-resonantly responds to electric oscillations at ~39 Hz [9,10]. Using an equivalent circuit model for MTs, we experi- mentally determine the capacitance and resistance of the MT network to be 1.27×10^{-5} F and 9.74×10^{4} Ω respectively, at physiological concentrations of tubulin. Our values indicate that the polymerisation of tubulin into MTs alters the spatial and temporal charge distribution, altering the electrical properties through charge storage in the cell.

2. Materials and Methods

2.1. Tubulin Reconstitution

Fluorescently labelled tubulin solution was prepared using previously published protocols [20]. Notably, tubulin stock powders that were devoid of MAPs were purchased. Lyophilised unlabelled tubulin powder (Cytoskeleton Inc, Denver, CO, USA; T240) was reconstituted in BRB80 supplemented with 1 mM GTP (guanosine triphosphate; Cytoskeleton Inc, Denver, CO, USA; BST06) and mixed with tubulin labelled with a rhodamine-based ester (Cytoskeleton Inc, Denver, CO, USA; TL590m) in a final labelling ratio of 1:15. Aliquots were snap-frozen and stored at −80 °C.

2.2. MT Polymerisation and Stabilisation

MT polymerisation was performed by incubating 45.45 µM tubulin aliquots in a 37 °C water bath for 30 minutes. BRB80 solution was heated alongside tubulin during the first 15 minutes of polymerisation. Subsequently, BRB80 was incubated at room temperature, and paclitaxel solution (Cytoskeleton Inc, Denver, CO, USA; TXD01; 2 mM stock) was thawed at room temperature alongside it. After 30 minutes of tubulin polymerisation brought to completion, 100 µL of BRB80 was added to 5 µL of 2 mM paclitaxel (BRB80T). For preparing 0.222, 2.225 and 22.225 µM MTs the above solution was added in different vol- umes to polymerised tubulin, as shown in Table 1. For preparing the BRB80T background for impedance measurements, 45 µL of this solution was added to 45 µL of BRB80.

Table 1

Volumes of tubulin and buffer solution (BRB80T or BRB80C) used to stabilise microtubules (BRB80T) or free tubulin (BRB80C) in solution.

Tubulin Concentration (μM)

0.222

2.222 22.225

Volume of BRB80T or BRB80C (μL)

99.5 95 5

Tubulin Volume (μL)

0.5 5 5

For tubulin stabilisation, 2 μL of 5 mM colchicine stock solution (Sigma-Aldrich, St. Louis, MO, USA; C9754; 5 mM in DMSO) was added to 100 μL BRB80 (BRB80C). Subsequently, a similar solution to BRB80T was prepared using colchicine. For preparing 0.222, 2.225 and 22.225 μM free tubulin solutions, the above solution was added in different volumes to free tubulin solutions, as shown in Table 1. For pre- paring BRB80C, 45 μL of this solution was added to 45 μL of BRB80.

2.3. Fluorescence Imaging of MTs

Imaging was performed on a Zeiss Examiner.Z1 microscope using a Hamamatsu (Hamamatsu City, Japan) EMCCD C9100 camera, a Zeiss (Oberkochen,

Germany) plan-Apochromat 1.4 NA 63x lens. After pipetting MT solution (2–5 µL) onto a glass slide (VWR 48382-173) a coverslip (VWR 48393-070) was placed on the solution, allowing it to spread. The microscope used an EXFO X-Cite 120 fluorescence source and excitation and emission filters of 535 nm and 610 nm, respectively. Exposure times between 50 ms and 300 ms were used for imaging to validate the presence of MTs.

2.4. Electrode Design and Device Construction

Each 'plate 'in the parallel-plate contact device was formed by FTO (fluorine-doped tin oxide)-coated glass slides (Sigma Aldrich, St. Louis, MO, USA; 735140). The slides were cleaved to dimensions of 1.5 mm × 10 mm × 50 mm for the upper contact and 1.5 mm × 27 mm × 50 mm for the lower contact. The cleaving dimensions were set using 3D printed devices that were placed as holders (The Shack, University of Alberta; Figure S1 in Supplementary Materials). The slides were ultrasonicated and subjected to reactive ion etching (RIE) using a 5-minute exposure to oxygen plasma (Oxford Instruments, Abingdon, UK; NGP80) to remove surface particulate matter. A 70-µm thick double-sided tape was used as a spacer, which formed a chamber of dimensions 3 mm × 1.25 cm × 70 µm. The top electrode was placed using a separate 3D-printed holder device (Figure S1). Once the device was constructed using the above protocol, solution was perfused into the chamber using a pipette and a filter paper for suction. We used flat copper electrode clips in a three-electrode configuration to connect to our capacitor device. The counter electrode was con- nected to the lower electrode, and the working and reference electrodes were connected to the top electrode of our device.

2.5. Impedance Measurements

Experiments were conducted using Electrochemical Impedance Spectroscopy (EIS) on a Zahner Zennium impedance analyser. The parallel-plate contact device was placed into the 3D-printed holder for stabilisation (Figure S1). The contacts from the machine were connected to the parallel-plate device using flat- faced copper alligator clips. A three-electrode configuration was used: The counter electrode was attached to the lower contact of the parallel-plate device, whereas the working electrode was attached to the upper contact with the reference electrode orthogonally clipped onto the clip of the working electrode. Within the Thales Z3.04 environment, the potentiostat mode was ON; the stabilisation delay was set to 1 s, the rest po- tential drift tolerance was set to 250 µV, V_{rms} was set to 5 mV. Solutions were perfused into the experimen- tal chamber using a micropipette tip at one opening, and a filter paper at the other opening for suction, sim- ilar to protocols used for Total Internal Reflection Fluorescence (TIRF) microscopy [31]. The frequency range of the EIS measurement was set from 4 MHz to 1 Hz and data were subsequently collected.

2.6. Data Analysis

MT and tubulin samples were analysed using data from five to seven days of experiments. Each day con- sisted of three to seven solutions for each concentration being tested, with one frequency sweep per solu- tion. Readings of each sweep were saved as a csv file, and next sample was loaded by solution exchange method. Water was run as the first solution for each day of experiments. BRB80T was run prior to MT solutions, and BRB80C were run prior to the free tubulin containing solutions. MT- and free tubulin-containing solutions were run on separate days, in increasing order of concentration. MATLAB (The Mathworks; Natick, MA, USA) scripts were used for data analysis. Fitting to the real and imaginary components of impedance was performed using the function lsqnonlin. Initial guess values for the MT network resistance and capacitance were 10^5 F and 10^{-5} Ω, respectively, based on visual inspection of raw data. The initial guess values for the nominal series resistor, R_H, were set at 1.78, 0.6 and 0.4 Ω with tubulin concentrations of 0.222, 2.222 and 22.222 µM, respectively. The 95% confidence intervals were determined using the function nlparci. Error propagation was performed assuming no relationship between various days of data collection.

3. Results

3.1. Validation of Parallel-Plate Contact Device to Measure Dielectric Properties of Physiologically Relevant Ionic Solutions

To determine the differences in the dielectric properties of solution caused by the presence of MTs, we first aimed to create an electrode geometry that would be experimentally robust and easily modelled. We fabricated an FTO-coated parallel-plate contact device (Figure 1a), which allowed EIS using a solution-exchange method.

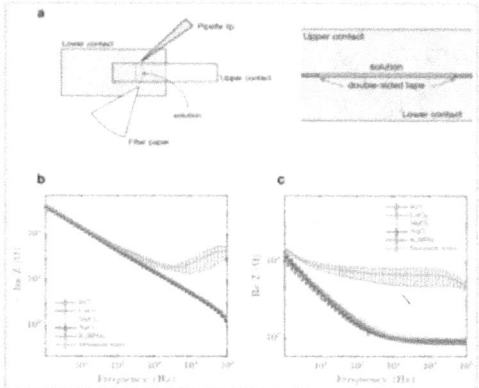

Figure 1

A parallel-plate contact device to measure the impedance properties of microtubules (MTs) compared to tubulin. The operation of the parallel plate device showing (**a**) top view (left) and side view (right). The upper and lower contacts, double-sided tape and solution are labelled in green, grey and blue, respectively. (**b**) Imaginary component of impedance for electrolytic solutions at 100 mM and de-ionised water. (**c**) Real component of impedance for electrolytic solutions at 100 mM and de-ionised water. Data display average values collected between 15 and 21 times. Error bars represent standard deviation.

We started by performing EIS on electrolytes found in the cytosol and observed that the imaginary component of impedance became less negative as a function of applied input frequency (Figure 1b). The total impedance of our system was given by:

$$Z = r_c + r_s/(1 + (r_s \omega C)^2) + j(\omega L_c - (r_s^2 \omega C)/(1 + (r_s \omega C)^2)),$$

Here, Z is the impedance, ω is the angular frequency (given by $2\pi f$ where f is the input voltage frequency), C is the system capacitance, L_c is the cable inductance, r_s and r_c are the solution and cable resistances respectively. We also observed a decrease in the real component of impedance as a function of input frequency (Figure

1c). Such a trend is expected from Warburg impedance [32,33] and is in accordance with the equation:

$$Z_{complex} = (A_\omega)/\sqrt{\omega} + (A_\omega)/(j\sqrt{\omega}),$$

Here, $Z_{complex}$ is the complex impedance and A_ω is the Warburg coefficient. Our circuit simplifies to the equation below if we ignore the effect of cable inductance ωL_c, at frequencies below 10^5 Hz:

$Z = r_c - j/\omega C$, Our results using various electrolytes emulated previous data [34,35,36] and validated the experimental

setup for further analysis.

3.2. The Effect of Microtubule Networks on Solution Capacitance at Physiologically Relevant Conditions

We reconstituted and polymerised fluorescent tubulin from a stock of 45.45 μM tubulin solution (Materials and Methods). MTs were stabilised using 50 μM paclitaxel [37,38] and imaged using an epi-fluorescence microscope. On diluting MT concentration across three orders of magnitude (0.222, 2.222 and 22.225 μM tubulin), we observed that while individual MTs at low concentrations were separated by large distances, those at cellular concentrations formed enmeshed networks reported previously (Figure 2a–c) [39]. Such interconnected MT networks are utilised by molecular motors for long-range macromolecular transport [40,41]. Here, their presence demonstrated successful MT polymerisation for electrical characterisation.

Figure 2

Microtubule imaging at different tubulin concentrations. Polymerisation was performed using 45 μM tubulin, and MTs were stabilised with 50 μM paclitaxel, and subsequently diluted to a final concentration of (**a**) 0.222 μM tubulin (**b**) 2.222 μM tubulin (**c**) 22.225 μM tubulin, respectively. Scale bars represent 10 μm.

We performed EIS on BRB80, BRB80T (BRB80 supplemented with 50 μM paclitaxel; background for all MT-containing solutions), and MT-containing solutions in increasing order of concentration (Figure 3a,b). We subtracted impedance values obtained for BRB80T alone from those in MT-containing solutions to de- termine the MT contribution to impedance. Our results showed that with an increasing MT concentration, the value of imaginary impedance became more negative, resulting in positive impedance differences (Figure 3c–f). This effect was greatest at the cell-like 22.225 μM tubulin concentration, with impedance dif- ferences lowering in magnitude with increasing input frequency (Figure 5a). Experiments with unpoly-

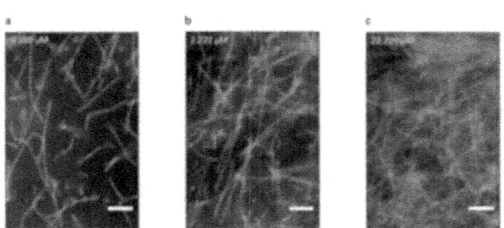

merised tubulin at the same concentrations were performed using the identical procedure, but using BRB80C (BRB80 was supplemented with 50 μM colchicine) as a background, to prevent MT nucleation [42,43]. Results with unpolymerised tubulin did not show an appreciable deviation from zero at any concentration (Figure S5a). The above results suggest that polymerisation of tubulin into MTs alters their ensemble electrical properties, increasing the solution's capacitance on forming MTs and their networks. An increase in the solution's capacitance because of MTs has previously been modelled, [7,44,45] indicating an increase in charge storage as free tubulin polymerises.

Figure 3

Examples of raw values of imaginary component of impedance in solutions containing (a) MTs and (b) tubulin, for the pur- pose of displaying typical impedance values. Mean differences in the imaginary component of impedance as a function of input AC (alternating current) frequency at total tubulin concentrations of (c) 22.225 μM (n = 22 experiments for tubulin, n = 21 for MTs), (d) 2.222 μM (n = 35 experiments for tubulin, n = 49 for MTs) (e) 0.222 μM (n = 35 experiments for tubu- lin, n = 49 for MTs), (f) comparison of the effects of paclitaxel (BRB80T) and colchicine (BRB80C, n = 49 experiments for BRB80T, n = 35 for BRB80C, n = 84 experiments for BRB80). Error-bars represent standard deviation.

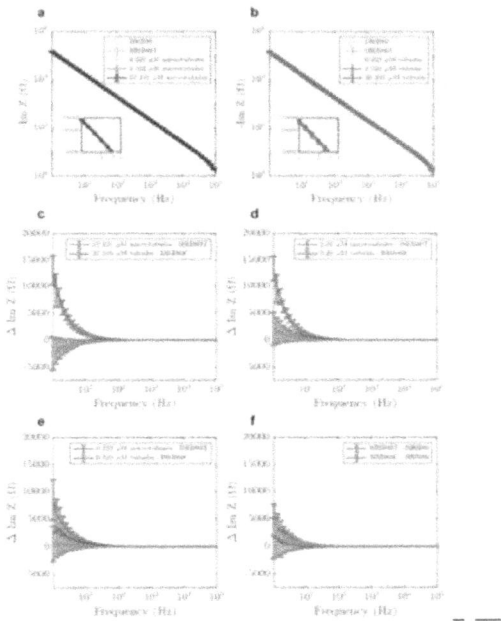

3.3. The Effect of Microtubule Networks on Solution Resistance at Physiologically Relevant Conditions

Next, we investigated the differences between MTs and tubulin in the real component of impedance (solution resistance). Previous studies using nanomolar tubulin concentrations and low ionic strengths (1–12 mM) have indicated that MTs enhance charge-transport in solutions [13,20,46]. To evaluate if this observa- tion held true at physiologically relevant tubulin concentrations and at higher ionic strengths, we also an- alysed the real component of impedance. Addition of both MTs and tubulin generally led to an increase in solution resistance (Figure 4a–f), with MTs having a higher resistance at low frequencies (1–20 Hz) com- pared to unpolymerised tubulin. Unexpectedly, a 'reversal 'of this behaviour was observed at higher frequencies as MTs began to lower the solution resistance compared to tubulin (Figure 5b). The reversal took place gradually between 10 and 300 Hz (Figure 6a–d), with a peak between 20 and 60 Hz (Figure 6e). Interestingly, within this range, we also found that the addition of MTs lowered solution resistance com- pared to background buffer BRB80T.

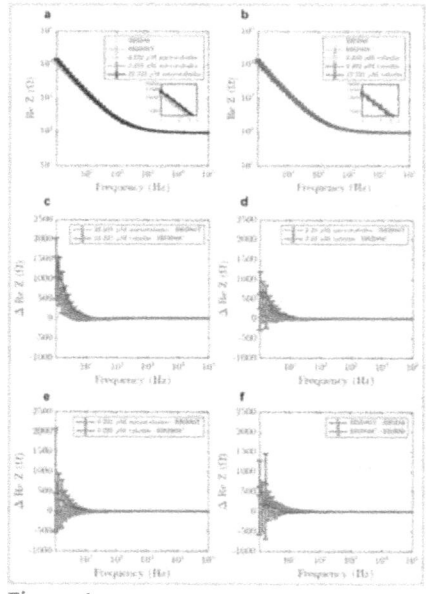

Figure 4

Examples of raw values of real component of impedance in solutions containing (**a**) MTs and (**b**) tubulin, for the purpose of displaying typical impedance values. Mean differences in the real component of impedance as a function of input AC fre- quency at total tubulin concentrations of (**c**) 22.225 µM (n = 22 experiments for tubulin, n = 21 for MTs), (**d**) 2.222 µM (n = 35 experiments for tubulin, n = 49 for MTs) (**e**) 0.222 µM (n = 35 experiments for tubulin, n = 49 for MTs), (**f**) compari- son of the effects of paclitaxel (BRB80T) and colchicine (BRB80C, n = 49 experiments for BRB80T, n = 35 for BRB80C, n = 84 experiments for BRB80). Error-bars represent standard deviation.

https://www.ncbi.nlm.nih.gov/pmc/articles/PMC7075204/

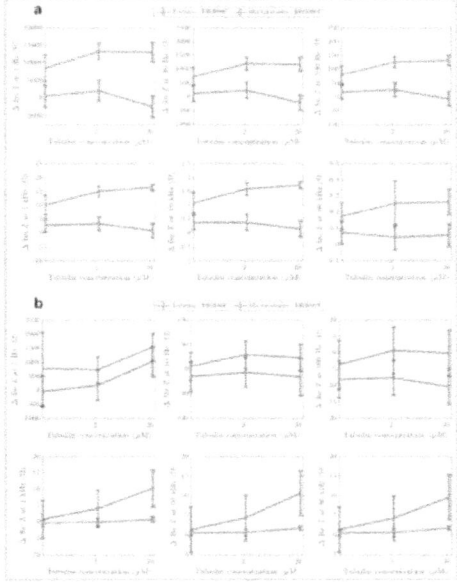

Figure 5

Graphs showing differences in the (**a**) imaginary from Figure 3 and (**b**) real component of impedance from Figure 4 as a function of tubulin concentration at input AC frequencies of 1 Hz, 10 Hz, 100 Hz, 1 kHz, 10 kHz and 86 kHz. Graphs dis- play average values. Error-bars represent standard deviation.

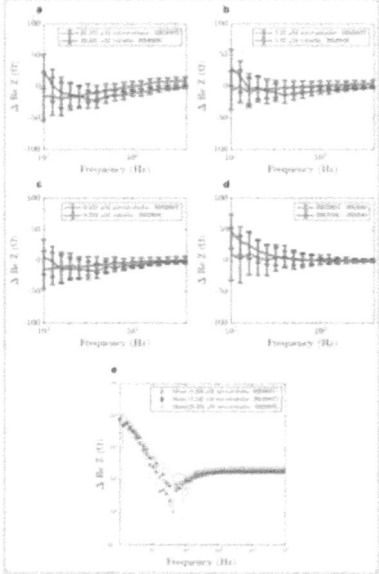

Figure 6

Zoomed in view of the mean differences in the real component of impedance as a function of decreasing input AC frequen- cy at total tubulin concentrations of (**a**) 22.225 µM, (**b**) 2.222 µM, (**c**) 0.222 µM, (**d**) comparison of the effect of paclitaxel and colchicine on impedance. (**e**) A logarithmic plot obtained by translating the graphs (a), (b) and (c) upwards. The trans- lation is performed by adding (1+minimum MT solution resistance) to the resistance of each MT concentration. A resis- tance reversal between 20–60 Hz is observed, with a peak at 39 Hz for the 22.225 µM concentration. Error-bars represent standard deviation.

Such a reversal in resistance between microtubules and tubulin has not been reported before. Because the extent of this reversal decreased with decreasing concentration, this result also displays the utility of our 'cell-like' approach. Our results are consistent with predictions of an increase in solution conductance at ~39 Hz [9,10], which have been hypothesised to arise from oscillatory ionic movement across the MT lat- tice through nanopores formed between adjacent tubulin dimers (Figure 9a).

It is worth noting that this region falls within the gamma frequency regime (20–60 Hz), implicating such quasi-resonant phenomena as a possible explanation for the source of low frequency intraneuronal electrical oscillations. No such reversal was observed for the corresponding frequency range in the imaginary im-

pedance values (Figure S4).

3.4. The Microtubule Network as an RC Circuit in Parallel

Our next aim is to quantify the resistance and capacitance of the microtubule network. The slope of approximately negative unity on the impedance difference curve suggested that the microtubule network re- sulted in the addition of a capacitive element to the solution. We examined several combinations but a par- allel RC (resistor-capacitor) circuit to represent the entire MT network provided the best fit to observed curves.

We modelled the impedance caused by external circuit elements and BRB80T as Z_o and Z_s respectively, as shown in Figure 7. The net impedance of the background BRB80T was thus given by:

$Z_{buffer} = Z_0 + Z_s$, (4a) Denoting the impedance, resistance and capacitance of the entire MT network by Z_{MT}, R_{MT} and C_{MT} respectively, the impedance for the circuit with MTs is given by: $Z_{MT+buffer} = Z_0 + Z_s + R_H + Z_{MT}$

where,

$1/Z_{MT} = 1/R_{MT} + j\omega C_{MT}$

Additionally, the impedance differences between solutions with and without MTs are given by:

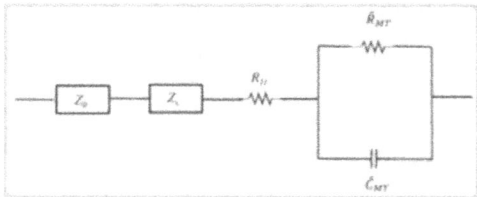

where

$\Delta Z = Z_{MT+buffer} - Z_{buffer} = R_H + Z_{MT}$,

$Z_{MT} = R_{MT}/(1 + (\omega C_{MT}R_{MT})^2) - j(\omega C_{MT}R_{MT}^2)/(1 + (\omega C_{MT}R_{MT})^2)$,

https://www.ncbi.nlm.nih.gov/pmc/articles/PMC7075204/

Figure 7

The equivalent electrical circuit model representing the microtubule network as a parallel RC circuit, with network resistance R_{MT} and capacitance C_{MT}. The external element has impedance Z_0, while solution has impedance Z_s. R_H is the small constant resistance that is ascribed to small fraction of unpolymerised tubulin that is present in MT containing solutions.

We subsequently fit experimental impedance difference curves shown in Figure 3 and Figure 4 to real and absolute value of imaginary parts of ΔZ using R_H, R_{MT} and C_{MT} as our fit parameters. Here, R_H is a resistance ascribed to the nominal fraction of unpolymerised tubulin present in MT containing solutions. The fitted curves are displayed in Figure 8 and the optimal fit parameters are listed in Table 2 (see Materials and Methods for details).

Figure 8

Mean differences of (**a**) imaginary and (**b**) real impedance curves for 0.222 µM, 2.222 µM and 22.222 µM, are fitted with the model described in Equation (6) and Figure 7. Fit parameters and confidence intervals are displayed in Table 2. The re- gion between 1–100 Hz was not fit because of the negative differences in resistance from background BRB80T solutions.

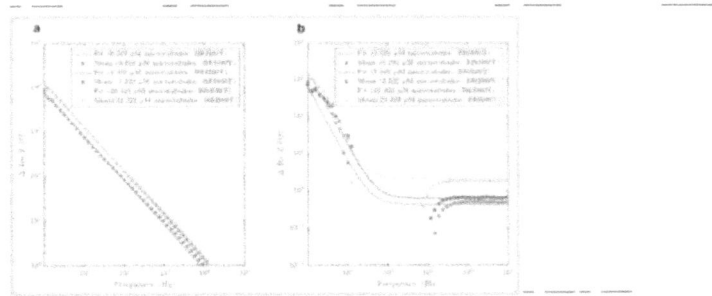

https://www.ncbi.nlm.nih.gov/pmc/articles/PMC7075204/

Table 2

Fit parameters attained by fitting the real and imaginary components of impedance to Equation (7). Fit parameters represent effective capacitance C_{MT}, and resistance R_{MT} introduced into the solution through the addition of the MT network at dif- ferent concentrations. R_H is the small constant resistance that is ascribed to small fraction of unpolymerised tubulin that is present in MT-containing solutions. γ_R and γ_C describe the effective resistance and capacitance per unit volume introduced by the microtubule network. δR_{MT}, δC_{MT} and δR_H correspond to 95% confidence intervals for the fit parameters. The val- ues $\delta \gamma_R$ and $\delta \gamma_C$ correspond to the uncertainties in the resistance and capacitance per unit volume. Corresponding graphs are displayed in Figure 8.

[Tub] (μM).

22.222 2.222 0.222

4. Discussion

C_{MT} (F).

1.27×10^{-5}

1.25×10^{-5}

2.01×10^{-5}

δC_{MT} (F) $1.48 \times$

10^{-7}

1.67×10^{-7}

3.38×10^{-7}

R_{MT} (Ω)

9.74×10^{4}

1.00×10^{5}

9.97×10^{4}

δR_{MT}

(Ω)

1.18×10^{4}

1.40×10^4

2.82×10^4

$R_H (\Omega)$ 2.12 0.61 0.41

δR_H

(Ω)

40.61 34.79 31.95

γ_R

(Ω/L)

3.71×10^{10}

3.81×10^{10}

3.80×10^{10}

γ_C

(F/L)

7.65 4.76 4.83

$\delta \gamma_R$

(Ω/L)

4.49×10^9

5.33×10^9

1.07×10^{10}

$\delta \gamma_C$

(F/L)

0.056 0.063 0.12

Our measurements using a parallel plate contact

device reveal interesting electrical properties of MTs at physiological concentrations. Unlike studies exposing MT-containing solutions to non-uniform electric fields [12,13,14,20], our device allowed robust quantification of electrical impedance in the presence of spatially uniform electric fields. Our results show that the addition of the MT network mimics a parallel RC element placed in series with the high-ionic strength solution, with a nonlinear dependence on MT number. Unpolymerized tubulin did not alter the capacitance significantly, indicating changes in electrical properties of tubulin as it polymerizes.

4.1. The Physical Underpinnings of An Increased Capacitance

An increase in capacitance arises from dense counterion condensation on the MT surface. This has been extensively predicted and simulated to arise from a variety of sources [7,8,44,47,48]. First, the negative charge of the tubulin dimer attracts counterions in solution, leading to the presence of a double layer and depletion region outside the microtubule surface [7,8,46,48]. The charge distribution in the MT protein wall is also highly non-uniform, with the outer surface containing approximately four times the charge compared to the inner surface [47] (Figure 9c). This asymmetry between the inner and outer electrostatic potentials serves to enhance capacitance and is responsible for the abnormally large dipole moment of the tubulin dimer [1]. The asymmetry also manifests through C-terminal 'tails 'composed of 10–12 amino-acids, that can extend 4–5 nm outwards from each tubulin monomer. These slender C-termini tails are highly negative, containing about 50% of the charge of the tubulin dimer [49]. As they stretch outwards into the solution in a pH and ionic strength-dependant manner, they increase the effective area of the tubu- lin dimer and significantly contribute to the overall MT capacitance [7,8].

Figure 9

Schematic of charge transport along and across an MT. (**a**) A representation of charge flow across the MT cross section through nanopores present between adjacent protofilaments. (**b**) A representation of charge flow through both inner and outer modes along an MT. Arrows depict charge flow via both mechanisms, enabling MT charge storage across a broad spectrum of frequencies, and charge transport at low AC frequencies in the cell. (**c**) Side view (left) and top view (right) of the tubulin dimer, displaying distribution of electrostatic potential at different locations. The negatively charged C-termini face towards the solution and contains ~50% of the total negative charge on a tubulin dimer.

Coherent oscillations of these C-terminal tails are modelled to generate solitonic pulses of mobile charge along the outer surface of an MT, creating ionic currents along its length [7,44,50]. Ions from the bulk so- lution are also modelled to be pumped into the hollow MT lumen through nanopores in its wall, resulting in charge accumulation inside the cylindrical MT over time [45]. A recent study using molecular dynamics simulations showed that the permeability of the MT lumen was significantly higher for Na^+ and K^+ as op- posed to Ca^{2+}, allowing for free movement of selective ions into the MT lumen across its porous surface [47]. To the best of our knowledge, our findings are the first to experimentally quantify this resistance en- countered by charge flow across the MT cross section. These results implicate not only ionic movement along the microtubule axis, but also across and inside it, enhancing the modelled roles of MTs as complex subcellular nanowires.

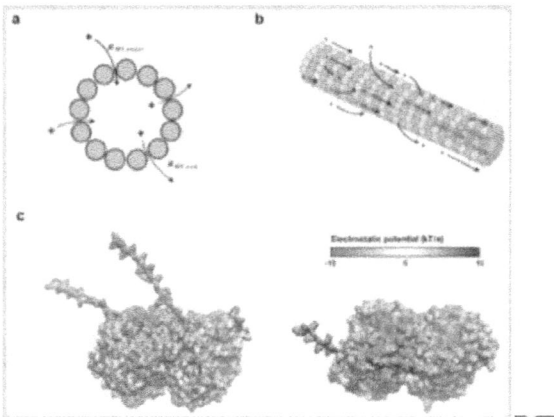

Manning's theory of polyelectrolyte solutions predicts the conditions for ionic condensation on charged polymer surfaces provided a sufficiently high linear charge density is present on these surfaces creating an ionic concentration depletion area surrounding them [51]. The sum total of the charges on polymer sur- faces and the associated counterions decreases to values dependent on the valence of the counterions and the Bjerrum length, which is the distance from the polymer surface at which the Coulomb energy of the screened surface charges equals the thermal energy. The double layer of surface charges and counterions separated by the Bjerrum length can be viewed as having capacitor-like properties. Although the Manning theory was originally developed for such polyelectrolytes as DNA, it was also applied to actin filaments [52] and MTs [53]. For actin filaments, its application explained the observed lossless transmission of elec- tric pulses along the filament lengths. In the case of MTs, it provided a plausible explanation of unusual amplification of injected electrical signals that propagated along these nanowires. The calculated Bjerrum length for MTs was found to be approximately 6.7×10^{-10} m [8,50]. Both actin filaments and MTs have been represented in these models by cable equations with effective real and imaginary impedance due to the viscosity of the solution-resisting ionic flows and the capacitive properties of the ionic double layers around the filaments, respectively [52,53]. The capacitance for a single ring of an MT including C-termini was calculated to be approximately 1.3×10^{-15} F [8]. When extended to 20 μm, (representative of the length of a single MT for our measurements), the predicted value would be $C = 3 \times 10^{-12}$ F, although an experimental confirmation of this prediction is not directly available through our measurements or in any previous work. We note the relatively weak dependence of network capacitance on MT concentration, and assign it to the random spatial locations and directional orientations of MTs in our solution. Indeed, the conductivity of randomly distributed RC networks has been shown to scale weakly with the number of elements in the network [54]. Additionally, qualitative similarities can be found in the models of random resistor and capacitor networks with a frequency-dependent crossover for both conductance and impedance in these networks due to percolation-type conduction [55]. We intend to develop a quantitative model for our experimental observations in a subsequent publication.

4.2. Implications for the Cell

Our work, which utilizes cell-like tubulin and ionic concentrations for the first time, indicates a cellular role for microtubules as wires that store charge. Neuronal environments where MTs are spontaneously nu- cleated from free tubulin, such as growth cones, experience large capacitance changes over short bursts of time. This ability significantly impacts the action potentials that are known to depend strongly on the local charge distributions [56]. Additionally, ionic movement across the MT wall enhances their roles as attenua- tors of local cation distributions. In nonneuronal environments, transient ionic currents around a MT dur- ing mitosis could impact MT dynamics and potentially influence the chromosome segregation. Specifically, Ca^{2+} ion storage/flow about an MT triggers its depolymerisation, whereas waves of Mg^{2+} or lowering in the local pH (increasing H^+) leads to MT stabilisation [57,58]. The attraction of Zn^{2+} or Mn^{2+} ions in the vicinity leads to the formation of two-dimensional tubulin polymers [59,60]. Properties of the cytoplasm such as polarisability and relative permittivity get severely attenuated because of the presence of MTs in the vicinity. Because of the polymerisation state of tubulin-altering solution capacitance, our findings im- plicate a temporal evolution of capacitance and ionic flows as the ratio of MTs to free unpolymerised tubu- lin changes [61,62,63]. MT lattice defects, which occur when a tubulin dimer is missing in an MT wall [64,65], cause a large ionic flux to develop at the defect site. Spatiotemporal charge distribution shifts are also critical at the MT end, where fluxes form because of sudden changes due to the polymerisation/de- polymerisation of the MT. Free/polymerised tubulin hence regulates local and global electrical properties,

https://www.ncbi.nlm.nih.gov/pmc/articles/PMC7075204/

creating spatially dynamic gradients of charge storage and flux. We envision a cytoskeleton that, in addi- tion to transporting macromolecules, stores and transports ionic signals and electrical information across the cytoplasm (Figure 9a,b).

Our findings can be coupled with a vast array of bio-nanodevices that utilises MTs and MAPs (micro- tubule-associated proteins) for construction of bio-nanotransporters and bio-actuators [66,67,68,69,70]. Under specific conditions, MAP-MT systems are capable of repositioning macromolecules [71,72], direc- tionally transporting microtubules [15,73] and even drive their movement within zero-mode waveguides [74] and inorganic nanotubes [75]. Storage of electrical charge and its transport along MTs can be coupled to such cutting-edge mechanical MAP-based devices to develop a wide range of nano-actuators and nano- sensors.

When compared to cells, the rates of MT nucleation and polymerisation are significantly lower in BRB80. This difference can be attributed to the absence of MAPs and macromolecular crowding [76,77]. Mammalian cells contain high concentrations of K^+ ions (140–300 mM) [27,28], which, in addition to MAPs and molecular crowding agents, will be included in a future study to attain physiological equiva- lence. We also note that the effect of PTMs (post-translational modifications) on the electrical properties of microtubules has not yet been explored. PTMs involve the addition of residues such as phosphate and glu- tamate that locally influence counterionic condensation around the outer microtubule surface.

We are in the process of performing DC (direct-current) measurements, determine the contribution of MTs to impedance relaxation time and evaluate the voltage dependence of capacitance on MT-containing solu- tions. Interestingly, this aspect has been discussed previously: the inductance of a single protofilament is calculated to be <1 fH [8]. Further investigation is required to experimentally confirm these predictions.

5. Conclusions

We used EIS to compare the complex impedance of MT- and tubulin-containing solutions. A physiologically relevant, high ionic strength buffer (BRB80) created a high noise, low impedance background, which was countered through the use of physiological concentrations of tubulin. While the presence of MTs increased solution capacitance, unpolymerised tubulin did not have any appreciable effect. In a study that is the first of its kind to the best of our knowledge, we determined the capacitance and resistance of the MT network at physiological tubulin concentrations to be 1.27×10^{-5} F and 9.74×10^{4} Ω. These values corre- spond to an effective resistance per unit volume of 3.71×10^{10} Ω/L and effective capacitance per unit vol- ume of 7.65 F/L. We envision a dual electrical role for MTs in the cell, that of charge storage devices across a broad frequency spectrum (acting as storage locations for ions), and of charge transporters (bio- nanowires) in the frequency region between 20 and 60 Hz. Our findings also indicate that the electrical properties of tubulin dimers change as they polymerise, revealing the potential impact of MT nucleation and polymerisation on the cellular charge distribution. Our work shows that by storing charge and attenuat- ing local ion distributions, microtubules play a crucial role in governing the bioelectric properties of the cell.

POST GRAD

Microtubule motor transport in the delivery of melanosomes to the actin-rich apical domain of the retinal pigment epithelium

Mei Jiang,[1,2,*] Antonio E. Paniagua,[1,2] Stefanie Volland,[1,2] Hongxing Wang,[1,2] Adarsh Balaji,[1,2] David G. Li,[1,2] Vanda S. Lopes,[1,2] Barry L. Burgess,[1,2] and David S. Williams[1,2,3,4,‡]

Author information Article notes Copyright and License information PMC Disclaimer

Associated Data

Supplementary Materials

Go to:

ABSTRACT

Melanosomes are motile, light-absorbing organelles that are present in pigment cells of the skin and eye. It has been proposed that melanosome localization, in both skin melanocytes and the retinal pigment epithelium (RPE), involves melanosome capture from microtubule motors by an unconventional myosin, which dynamically tethers the melanosomes to actin filaments. Recent studies with melanocytes have questioned this cooperative capture model. Here, we test the model in RPE cells by imaging melanosomes associated with labeled actin filaments and microtubules, and by investigating the roles of different motor proteins. We found that a deficiency in cytoplasmic dynein phenocopies the lack of myosin-7a, in that melanosomes undergo fewer of the slow myosin-7a-dependent movements and are absent from the RPE apical domain. These results indicate that microtubule-based motility is required for the delivery of melanosomes to the actin-rich apical domain and support a capture mechanism that involves both microtubule and actin motors.

KEY WORDS: Melanosome, Retina, Dynein, Kinesin-1, Myosin-7a

Summary: The motility and localization of melanosomes in the retinal pigment epithelium are dependent on cytoplasmic dynein and myosin-7a, thus supporting a capture mechanism that involves both microtubules and actin motors.

Go to:

INTRODUCTION

Melanosomes are organelles that originate from endosomes and contain melanin pigments (Marks and Seabra, 2001). They are present in the skin and the eye. An important function of melanosomes is in the screening of light, which can be a dynamic process, effected by melanosome motility, in response to changes in ambient lighting. Because melanosomes can be imaged with standard bright-field microscopy, without overexpression of tagged protein markers, they represent an excellent system for investigating organelle motility and the molecular motor systems that drive the underlying transport.

In mammalian skin, melanosome biogenesis occurs in melanocytes (Raposo and Marks, 2007). The melanosomes then pass from the dendrites of the melanocytes to neighboring keratinocytes (Hume et al., 2001; Wasmeier et al., 2008; Fukuda, 2013; Wu and Hammer, 2014; Moreiras et al., 2019). Preceding the intercellular transfer, melanosomes must be transported and retained in the melanocyte dendrites. Retention of the melanosomes depends on myosin-5a. Early studies showed that the melanocyte dendrites in *Myo5a*-null mice (known as *dilute* mice because of their light coat color) do not contain melanosomes (Provance et al., 1996; Wei et al., 1997). It was then shown that myosin-5a is linked to melanosomes by RAB27A and a member of the exophilin family, melanophilin. Live-cell studies led to a model in which melanosomes are transported to the cell periphery by microtubule motors and then passed onto myosin-5a, which retains them in the dendrite (i.e. a cooperative capture model) (Wu et al., 1998; Desnos et al., 2007; Hammer and Sellers, 2011). The main microtubule motor driving the movement to the periphery is indicated to be the plus-end directed motor kinesin-1 (Hara et al., 2000; Vancoillie et al., 2000). More recently, this model has been challenged, based on the lack of requirement for microtubule integrity for delivery of melanosomes to the dendrites, and lack of enrichment of kinesin-1 on melanosomes. Instead, recent evidence supports processive myosin-5a transport entirely from the center of the melanocyte to its periphery (Evans et al., 2014; Robinson et al., 2017).

In the eyes of invertebrates and vertebrates, movements of melanosomes provide a means to alter visual sensitivity and resolution. Clear examples are found among mollusks (Daw and Pearlman, 1974), arthropods (Williams, 1982; Narendra et al., 2016) and fish and amphibians (Back et al., 1965; Burnside, 2001). In the retinal pigment epithelium (RPE) of vertebrate eyes, cylindrically shaped melanosomes enter the narrow apical processes that project among the photoreceptor outer segments (POSs), with their long axis parallel to the POSs and thus the direction of incoming light (Burnside et al., 1983; Burgoyne et al., 2015). In lower vertebrates, movement into the apical processes upon light exposure and withdrawal from them upon darkness are major events, affording significant changes in the light-guiding properties of the POSs. Although more muted, melanosomes have been observed to move into the RPE apical processes of mice in response to light onset (Futter et al., 2004). The effect of this movement is unclear, but it might be related to functions other than light absorption that have been identified for RPE melanosomes.

For example, melanosomes have been shown to have a cytoprotective effect in RPE cells under nonphotic oxidative stress (Burke et al., 2011). Melanosomes might also contribute to the enormous phagocytic load incurred by RPE cells; for example, in the mouse central retina each RPE cell is associated with 200 POSs (Volland et al., 2015) and, each day, peaking at light onset, the distal 10% of each POS is ingested and then degraded (Young, 1967; LaVail, 1976). RPE melanosomes contain proteases (Azarian et al., 2006), including cathepsin D, a major enzyme in the degradation of POS proteins (Hayasaka et al., 1975), and have been observed to fuse with phagosomes (Schraermeyer et al., 1999).

Although lack of myosin-5a does not affect the localization of RPE melanosomes (Gibbs et al., 2004), another unconventional myosin, myosin-7a, does. In humans, mutations in the gene encoding myosin-7a underlie Usher syndrome type 1B, a deaf-blindness disorder (Weil et al., 1995). One of several retinal functions for myosin-7a (Williams and Lopes, 2011) involves the apical localization of RPE melanosomes, including their presence in the apical microvilli (Liu et al., 1998). As for melanocyte melanosomes, RPE melanosomes require RAB27A for their localization but, instead of melanophilin, they require another exophilin (myosin VIIA and Rab interacting protein; MYRIP), which links RAB27A and myosin-7a (El-Amraoui et al., 2002; Futter et al., 2004; Gibbs et al., 2004; Kuroda and Fukuda, 2005; Klomp et al., 2007). Therefore, apical localization of RPE melanosomes is comparable to the dendritic localization of melanocyte melanosomes, with a RAB27A–MYRIP–myosin-7a complex used by the RPE in a similar way as the RAB27A–MLPH–myosin-5a complex is used by melanocytes. However, the motility of melanosomes in the RPE cell body has not been specifically studied, and it is not known how melanosomes move from the RPE cell body to the apical region. Thus, a cooperative capture model has not been tested.

Melanosomes in primary cultures of RPE cells lacking myosin-7a undergo a larger number of fast long-range movements than in similarly cultured wild-type RPE cells (Gibbs et al., 2004; Lopes et al., 2007a). It was suggested that these faster movements are driven by microtubule motors. In the present study, we test this proposal by direct observation of the motility of RPE melanosomes in the presence of labeled actin filaments and microtubules. In addition, we identify microtubule motors that associate with the melanosomes and are required for the normal motility and localization of melanosomes. We compare and contrast our findings with reports on melanosome motility in melanocytes.

Go to:

RESULTS

RPE melanosomes move along actin filaments

Previous studies have demonstrated that myosin-7a is required for normal melanosome localization (Liu et al., 1998) and motility dynamics (Gibbs et al., 2004; Lopes et al., 2007a) in RPE cells, suggesting that melanosomes are tethered to or move along actin filaments by this actin-based motor. We attempted to observe the interaction between melanosomes and actin filaments by direct imaging of live RPE cells isolated from wild-type mice and expressing RFP-actin or GFP-tractin; GFP-tractin labels only F-actin (Johnson and Schell, 2009; Yi et al., 2012). Imaging indicated that the melanosomes were associated with labeled actin filaments and were typically oriented with their long axis parallel to the filaments. Melanosomes were motionless for quite long periods, as if tethered to an actin filament, with occasional relatively short bursts of movement along the labeled actin filaments, with their long axis remaining in line with the direction of movement (Movies 1, 2). Fig. 1A shows the track of a melanosome along a labeled actin filament (or a bundle of parallel actin filaments), as demonstrated in Movie 1.

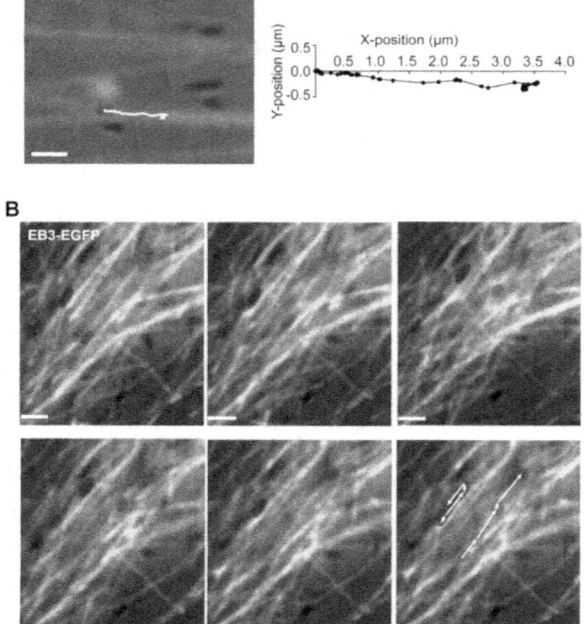

Fig. 1.
Movement of melanosomes along actin filaments and microtubules in mouse primary RPE cells. (A) Movement of melanosomes along actin filaments in a cultured primary mouse RPE cell, transduced with Cellular Lights Actin-RFP. Single frame from the start of **Movie 1**. The white line shows the track that a melanosome takes along an actin filament, evident in subsequent frames. The track is shown in detail on the right. (B) Frames from **Movie 3**, showing movement of melanosomes along EB3-EGFP labeled microtubules in RPE cells isolated from shaker1 mice (in this case, the *shaker1₄₆₂₆SB* allele, which is null for *Myo7a*). The blue pseudocolored melanosome moves first to the upper right, and then reverses course. The red pseudocolored melanosome undergoes a longer run towards the upper right. The complete tracks of the two melanosomes are shown in the last panel. The trajectories were obtained using Volocity software and are shown as a white line, with small arrowheads indicating the direction of movement. EB3-EGFP labels the microtubule growing (plus) ends, most of which (as shown in **Movie 3**) migrate to the lower left of the field of view. Scale bars: 2 μm.

RPE melanosomes move along microtubules

As a first test of whether the fast long-range movements of RPE melanosomes occur along microtubules, we imaged melanosomes in primary cultures of shaker1 RPE cells that expressed tubulin-GFP or EB3-EGFP in order to label the microtubules fluorescently. We studied cells from *shaker1₄₆₂₆SB* mice, which have a *Myo7a*-null mutation (Liu et al., 1999), in order to eliminate myosin-7a-based melanosome motility on actin filaments. Thus, we could isolate microtubule motility from myosin-7a-based motility on any (unlabeled) actin filaments that were close to the microtubules. Use of shaker1 RPE also increased the number of melanosomes available for microtubule movements (Gibbs et al., 2004; Lopes et al., 2007a).

Melanosomes were evident moving along the labeled microtubules (Fig. 1B; Movie 3). Compared with tracks along actin filaments in wild-type RPE cells (Fig. 1A; Movies 1, 2), this movement of melanosomes along microtubules occurred at faster speeds and with longer run lengths. Like the movement along actin filaments, the long axis of a melanosome remained aligned with the microtubule and thus the direction of movement. Movement in one direction was followed by at least one clear reversal of direction for 18% of the observed melanosomes during 3 min of imaging (see melanosome indicated by arrowhead in Fig. 1B). These bidirectional movements suggest the involvement of both plus-end and minus-end directed microtubule motors. Fast melanosome movements were also observed along labeled microtubules in wild-type RPE cells, although they occurred less frequently and over shorter distances.

Dependence on the presence of microtubules for melanosome motility has been shown previously; exposure of mouse primary RPE cells to 10 µm nocodazole for 1 h resulted in complete elimination of any motility in the majority of melanosomes (Lopes et al., 2007a). A similar finding was observed in the present study when comparing melanosome motility in untreated and nocodazole-treated mouse primary RPE cultures. The melanosomes that still moved in nocodazole-treated cells underwent only short-range movements at slow speeds (Fig. 2), of the type attributed to myosin-7a motility on actin filaments (Gibbs et al., 2004; Lopes et al., 2007a).

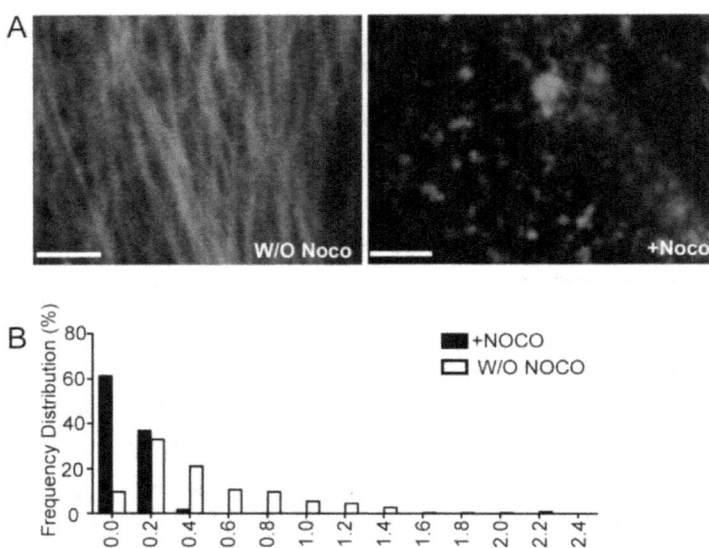

Fig. 2.
Effect of nocodazole disruption of microtubules on melanosome motility in RPE cells. (A) Primary wild-type mouse RPE cells were transfected with tubulin-mCherry and treated with 10 μM nocodazole (Noco). Cells were imaged before (left) and 30 min after (right) treatment with nocodazole. (B) Distribution of speeds of melanosomes that were motile in nocodazole-treated and untreated cultures; note that more than 80% of melanosomes were immotile following nocodazole treatment (and thus not counted). Melanosomes were imaged at 2 frames/s and speeds determined from the frame-to-frame displacement (also known as instantaneous speed). Speeds were binned, such that speeds <0.2 μm s-1 were collected under 0 μm s-1, speeds ≥0.2 but <0.4 μm s-1 were collected under 0.2 μm s-1, etc. Scale bars: 5 μm.

For the remainder of the study, we tested the

roles of kinesin-1 and cytoplasmic dynein in driving this microtubule motility.

Kinesin-1 involvement in RPE melanosome transport

Studies on kinesin-1 indicate that it functions in RPE phagosome motility (Jiang et al., 2015), although not in the motility of melanosomes in melanocytes (Robinson et al., 2017). To investigate whether kinesin-1 plays a role in RPE melanosome motility, we focused on its ubiquitous heavy chain, KIF5B (Xia et al., 1998). In support of a role, immuno-electron microscopy (immunoEM) of retinal sections showed that KIF5B was associated with melanosomes. Immunogold label was detected on melanosomes in all regions of the cell (**Fig. 3**), although quantification of the label (from 80 fields of view, aggregated from sections from three animals) showed that melanosomes in the cell body contained 2.2 times more label than those in the apical microvilli, indicating a somewhat higher association between KIF5B and melanosomes in the region containing microtubules (Jiang et al., 2015).

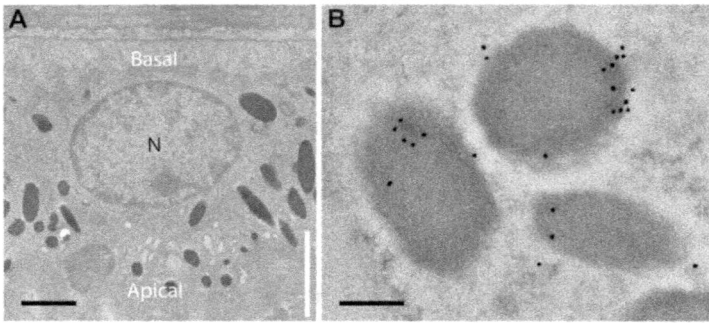

Fig. 3.
Immunogold labeling of melanosomes in mouse RPE cells by KIF5B antibodies. (A) EM of mouse RPE cell. The apical region of the cell includes the apical microvilli (region indicated by vertical white line). These microvilli interdigitate with the tips of photoreceptor outer segments. The actual boundary between the actin-rich apical domain and the cell body (which contains microtubules throughout) corresponds to the level of the junctional complexes and a circumferential ring of actin filaments, and so is slightly more basal than the base of the microvilli (Jiang et al., 2015). N, nucleus. (B) Example of melanosomes labeled by KIF5B immunogold in the cell body region. Scale bars: 1 μm (A), 200 nm (B).

To test for the involvement of KIF5B in melanosome transport, conditional knockout mice (Cui et al., 2011) were studied. Primary RPE cells isolated from *Kif5b*flox/flox mice were transfected with a Cre/GFP coexpression vector. Western blot analysis showed near complete depletion of KIF5B in the presence of Cre (Fig. 4A). Tracks of melanosome movements were obtained by imaging cells in primary RPE cultures from *Kif5bflox/flox* mouse littermates, with Cre expression (indicated by GFP) and without Cre expression. Analysis of the tracks showed that Cre expression resulted in a small shift in the distribution of measured maximal speeds, such that there was a lower frequency of slower maximal speeds (≤0.6 μm s-1) and a higher frequency of faster maximal speeds (≥0.8 μm s-1) (**Fig. 4B**).

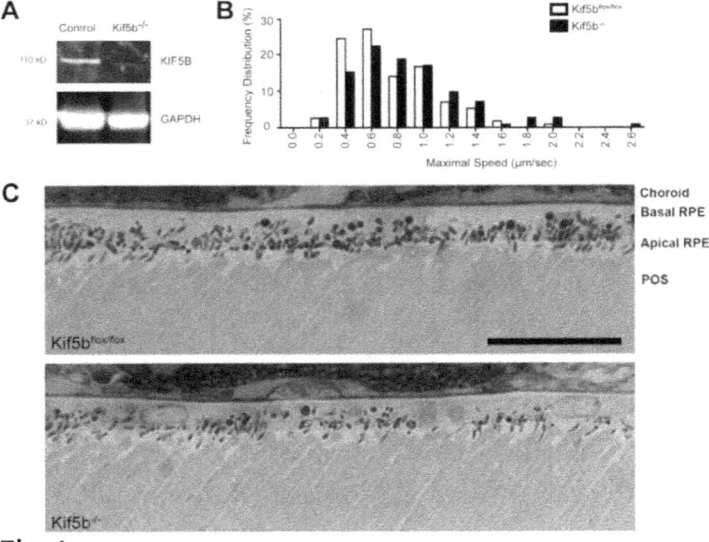

Fig. 4.
Effect of loss of KIF5B on melanosome speed and localization. (A) Western blot of RPE cells isolated from *Kif5bflox/flox* mice transfected with Cre (to generate *Kif5b-/-*) or without Cre (control). The blot was probed with antibodies against KIF5B and GAPDH (loading control). Apparent molecular masses

are indicated. (B) Frequency distribution of the maximal speed measured over 3 min intervals for melanosomes in *Kif5b*^{*flox/flox*} primary RPE cultures, with Cre (*Kif5b -/-*) or without Cre (*Kif5b*^{*flox/flox*}). Data were obtained from 3 min tracks ($n=56$ tracks for *Kif5b-/-*, $n=57$ tracks for *Kif5b*^{*flox/flox*}, aggregated from 3 separate experiments) of active melanosomes (i.e. with straight-line displacement of more than 2 μm after 3 min). Maximal speeds were binned, such that speeds <0.2 μm s-1 were collected under 0 μm s-1, speeds ≥0.2 but <0.4 μm s-1 were collected under 0.2 μm s-1, etc. (C) Light microscopy images of retinas from *Kif5b*^{*flox/flox*} mice (*Kif5b*^{*flox/flox*}) and *Kif5b*^{*flox/flox*}*;BEST1-cre* (*Kif5b -/-*) mouse retinas. The different retinal layers, choroid, basal RPE, apical RPE and the photoreceptor outer segments (POS), are indicated in the top panel. Scale bar: 25 μm.

These results suggest that KIF5B (and thus kinesin-1) contributes to the normal motility of melanosomes on microtubules. However, the change in motility was not associated with a significant change in the distribution of melanosomes in the RPE. The RPE in retinal sections from *Kif5b*^{*flox/flox*}*;BEST1-Cre* mice contained melanosomes in their apical microvilli as well as in the cell body, as in comparable sections from control mice (Fig. 4C).

Dynein involvement in RPE melanosome transport

To test for evidence of melanosome transport by cytoplasmic dynein, we first performed immunoEM with antibodies against both intermediate chains of cytoplasmic dynein 1 (DYNC1). Similar to the result obtained with KIF5B immunoEM, melanosomes were labeled (Fig. 5A) and the density of melanosome label in the basal RPE was 2.4 times that measured on melanosomes in the apical microvilli (quantified from 47 fields of view, aggregated from sections from three animals).

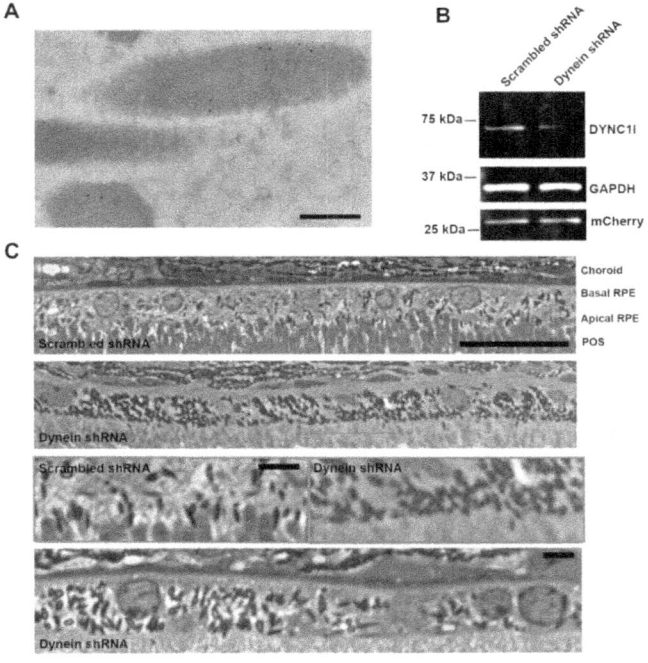

Fig. 5.
Immunogold labeling of dynein and analysis of the effect of dynein deficiency on melanosome localization. (A) Example of melanosomes in the cell body region, labeled by antibodies against both intermediate chains of cytoplasmic dynein 1. (B) Western blot of RPE cell cultures transduced with lentivirus containing scrambled shRNA-mCherry or cytoplasmic dynein heavy chain (DYNC1H1) shRNA-mCherry. The blot was probed with antibodies against cytoplasmic dynein 1 intermediate chains 1 and 2 (DYNC1I), GAPDH (loading control) and mCherry (transduction control). (C) Light microscopy images of RPE cells from wild-type mice, injected subretinally with scrambled shRNA or DYNC1H1 shRNA. The different retinal layers, choroid, basal RPE, apical RPE and the photoreceptor outer segments (POS), are indicated in the top panel. The regions framed by the red boxes (top two panels) are shown at higher magnification in the next panel. The image in the bottom panel is from another retina that was injected with DYNC1H1 shRNA. In the scrambled shRNA-treated retinas, melanosomes were present in the apical RPE, whereas in the DYNC1H1 shRNA-treated retinas, melanosomes were largely absent from the apical RPE. The shRNA plasmids were delivered to the RPE cells by injection into the subretinal space, followed by electroporation. The regions of the retina examined corresponded to those near the site of plasmid injection, and therefore their relative location varied somewhat from animal to animal. Note that the number of melanosomes per RPE cell varies across normal mouse and rat retinas, even though melanosomes are still present in the apical processes of all regions (**Howell et al., 1982; Williams et al., 1985**). Scale bars: 200 nm (A), 25 μm (C, top two and bottom panels), 10 μm (C, third panel).

Next, we used dynein heavy chain (DYNC1H1) shRNA, as described and characterized previously in a study of axonal transport in mouse neurons (Encalada et al., 2011), to investigate the effect of reducing the amount of DYNC1 heavy chain. We tested the efficacy of this shRNA in reducing DYNC1 intermediate chain expression using lentiviral transduction of primary RPE cultures, so that the majority of cells expressed the shRNA plasmids. Fig. 5B illustrates significant depletion of DYNC1 intermediate chains with the DYNC1H1 shRNA, in contrast to no effect with a scrambled shRNA.

Using subretinal injection and electroporation, we were able to transfect RPE cells with DYNC1H1 shRNA, and thus knock down RPE DYNC1 activity *in vivo*. In the areas of the retina transfected with DYNC1H1 shRNA, melanosomes were absent from the apical RPE; they did not surround the tips of the photoreceptor outer segments as in RPE that was not transfected or was transfected with scrambled shRNA (Fig. 5C). Although the melanosomes were absent from the apical RPE, they did not congregate in a particular region of the cell body; they appeared to be distributed throughout the cell body, like melanosomes in wild-type RPE, and, indeed, like melanosomes in *Myo7a*-mutant mice (Liu et al., 1998).

To test how loss of DYNC1 affects melanosome motility, we tracked melanosomes in primary RPE cell cultures transfected with DYNC1H1 shRNA or scrambled shRNA (as a control). Treated cells in the culture could be identified by expression of mCherry. DYNC1H1 shRNA shifted the distribution of maximal melanosome speeds, such that there was a lower frequency of slow maximal speeds (<0.4 µm s^{-1}) and a higher frequency of faster maximal speeds (≥0.6 µm s^{-1}), compared with the effect of scrambled shRNA (Fig. 6A). The extent of this change in the distribution of maximal speeds was much more marked than that observed with the deletion of KIF5B; it was comparable to that observed in shaker1 mice as a result of loss of myosin-7a function (Gibbs et al., 2004; Lopes et al., 2007a).

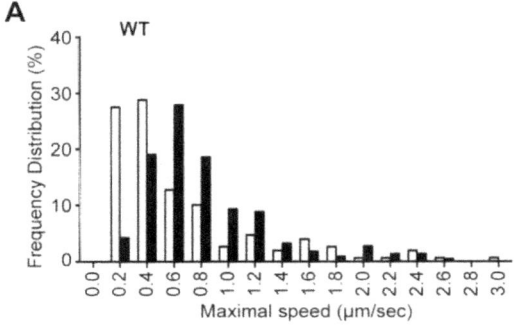

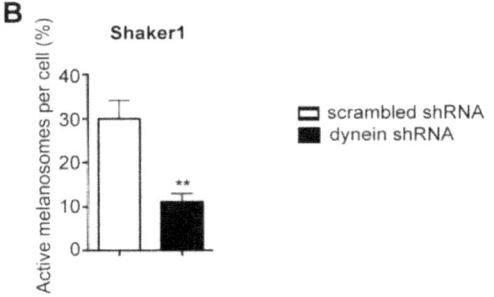

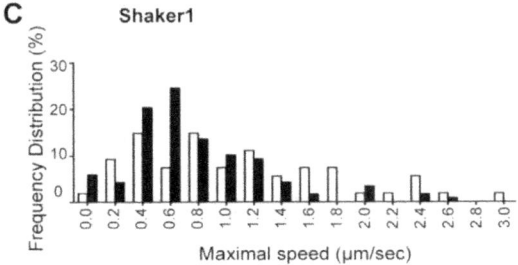

Fig. 6.
Effect of loss of dynein on melanosome motility. (A–C) White bars, scrambled shRNA; black bars, DYNC1H1 shRNA. (A) Frequency distribution of the maximal speed measured for melanosomes in primary cultures of RPE cells treated with scrambled (control) shRNA or DYNC1H1 shRNA. Data were obtained from 3 min

tracks ($n=75$ tracks for scrambled shRNA, $n=106$ tracks for DYNC1H1 shRNA, aggregated from three separate experiments) of active melanosomes (i.e. with straight-line displacement of more than 2 μm after 3 min). Maximal speeds were binned, such that speeds <0.2 μm s-1 were collected under 0 μm s-1, speeds ≥0.2 but <0.4 μm s-1 were collected under 0.2 μm s-1, etc. The percentage of active melanosomes, determined from three independent cell cultures, was similar for cells treated with scrambled shRNA (15±3% s.e.m.) and cells treated with DYNC1H1 shRNA (14±3% s.e.m.). (B) The percentage of active melanosomes (i.e. with straight-line displacement of more than 2 μm after 3 min) in shaker1 primary RPE cells, treated with scrambled shRNA or DYNC1H1 shRNA. Data were obtained from four separate experiments; **$P=0.0007$ (Mann–Whitney test, two-tailed). Error bars indicate s.e.m. (C) Frequency distribution of the maximal speed measured for melanosomes in primary cultures of shaker1 RPE cells treated with scrambled shRNA or DYNC1H1 shRNA. Data were obtained from 3 min tracks ($n=27$ tracks for scrambled control shRNA, $n=60$ tracks for dynein shRNA, aggregated from three separate experiments) of active melanosomes (i.e. with straight-line displacement of more than 2 μm after 3 min). Maximal speeds were binned as in A.

We examined the effects of DYNC1 knockdown further by studying shaker1 RPE, so that melanosomes would not be recruited to actin filaments by myosin-7a, thus helping isolate motility on microtubules. As noted, loss of myosin-7a results in faster speeds of the motile melanosomes (Gibbs et al., 2004; Lopes et al., 2007a). This effect can be seen by comparing the control (scrambled shRNA) bars in **Fig. 6A** with those in **Fig. 6C**. In shaker1 RPE, DYNC1 knockdown resulted in reduction of the proportion of motile melanosomes from 30% to just 10% (**Fig. 6B**). Interestingly, DYNC1 knockdown did not significantly alter the distribution of maximal speeds of those melanosomes that were motile (**Fig. 6C**). It appears that the shift to a higher frequency of faster speeds with DYNC1 knockdown is a result of reduced delivery of melanosomes to the actin filaments and reduced myosin-7a motility, consistent with the observations of retinal sections (**Fig. 5C**).

Go to:

DISCUSSION

We report direct imaging of RPE melanosomes moving along labeled actin filaments and microtubules. The long axis of the melanosome is aligned with the actin filament or microtubule, and thus the direction of movement. Most melanosomes were paused on the actin filaments, as if tethered, and underwent relatively short movements. Movements along microtubules were faster, involved longer runs and were sometimes bidirectional. In addition, we investigated the roles of kinesin-1 and cytoplasmic dynein in driving melanosome motility along microtubules. ImmunoEM demonstrated that both motors were associated with melanosomes. Knockout of the kinesin-1 heavy chain gene, *Kif5b*, had a relatively mild effect, with no significant alteration in melanosome localization. However, shRNA knockdown of DYNC1H1 completely phenocopied lack of myosin-7a, in terms of melanosome motility in RPE cell culture and melanosome localization *in vivo*, indicating that cytoplasmic dynein is essential for the passage of melanosomes from the cell body to the actin-rich apical domain of the RPE.

The organization of microtubules in differentiated RPE cells is comparable to that of other epithelial cells (Bacallao et al., 1989; Gilbert et al., 1991), with some lateral microtubules emerging from an apical centrosome in the apical part of the cell body and some vertical microtubules, which are not associated with a centrosome. The majority (71%) of the vertical microtubules are oriented with their plus ends more basal (Jiang et al., 2015), so that minus-end directed movements direct melanosomes predominantly towards the RPE apical domain. A deficiency in dynein resulted in the same dislocalization of melanosomes from the apical RPE, and the same increase in motility, as observed with shaker1 RPE (Liu et al., 1998; Gibbs et al., 2004). Only a small minority of RPE cells contained any melanosomes in the actin-rich apical domain; perhaps DYNC1H1 was incompletely knocked down in these cells. In live RPE cells deficient in dynein, the distribution of maximal speeds was shifted away from the slower speeds that have been attributed to myosin-7a driven motility (Udovichenko et al., 2002). These findings indicate the importance of delivery of melanosomes to the minus-ends of the microtubules by dynein, so that they can be transferred from the cell body to the RAB27A–MYRIP–myosin-7a complex in the actin-rich apical domain. Thus, they support a cooperative capture model.

Despite evidence of some bidirectional movement of melanosomes on microtubules, and an effect on melanosome motility in response to loss of KIF5B, kinesin-1 had no obvious effect on melanosome localization. Loss of KIF5B resulted in a slight shift to faster maximal speeds (Fig. 4B). As with loss of DYNC1H1, this shift suggests fewer movements on actin filaments. However, the extent of the shift was much less than that observed with knockdown of dynein. Perhaps this milder phenotype is due to the involvement of another plus-end microtubule motor that normally functions with kinesin-1 in melanosome motility. The shift to faster maximal speeds in the absence of KIF5B suggests that such an additional motor mediates more rapid movements, which are inhibited in the presence of kinesin-1. Kinesin-1 and kinesin-3 have both been found to contribute to plus-end motility of cargoes in neuronal axons and HeLa cells (Guardia et al., 2016; Lim et al., 2017). Similarly, kinesin-1 and heterotrimeric kinesin-2 are both involved in the axonal transport of acetylcholine esterase vesicles (Kulkarni et al., 2017). Heterotrimeric kinesin-2 appears to be a good candidate to partner in melanosome motility, given its role in melanosome transport in *Xenopus* melanophores (Tuma et al., 1998).

An alternative possibility is that the absence of kinesin-1 results in more dynein motility. However, the *in vivo* speed of cytoplasmic dynein has been indicated to be slower than that of kinesin-1 (Howard, 2001), in which case, the observed shift to faster speeds seems inconsistent with increased dynein activity. Moreover, more dynein activity might be expected to be associated with a greater apical localization of melanosomes, which we did not detect.

Skin melanocytes and RPE cells are two mammalian cell types in which melanosomes move from the cell body to the cell periphery. In melanocytes, melanosome biogenesis is active (Raposo and Marks, 2007) and the melanosomes do not return to the cell body; they are delivered to keratinocytes from the melanocyte dendrites (Hume et al., 2001; Wasmeier et al., 2008; Fukuda, 2013; Wu and Hammer, 2014; Moreiras et al., 2019). By contrast, melanosome biogenesis is very low in the postnatal RPE (Lopes et al., 2007b). Another difference between the two cell types is in the cytoskeletal organization. With the epithelial polarity and microtubule organization of RPEs (Jiang et al., 2015), movement towards the minus-ends of most microtubules directs melanosomes towards the actin-rich apical domain, where they can be captured by myosin-7a. In melanocytes, dynein moves melanosomes towards the cell nucleus, away from the dendrites (Byers et al., 2000), so that a plus-end microtubule motor is needed for movement towards the dendrites. It has been proposed that kinesin-1 fulfills this role, and is recruited to melanosomes by RAB1A (Ishida et al., 2012, 2015). However, more recently, it has been argued that KIF5B and RAB1A are not recruited to melanocyte melanosomes to a sufficient extent, and that the main driving force for the dendritic delivery of melanosomes comes from myosin-5a moving along cortical actin filaments (Evans et al., 2014; Robinson et al., 2017). In this case, melanocytes and RPE cells would differ, with a cooperative capture mechanism for melanosome delivery only present in RPE cells. However, consistent with the observations made here on kinesin-1 in RPEs, perhaps kinesin-1, together with another kinesin, functions in melanosome transport to the melanocyte dendrites and thus provides additional microtubule-based driving force.

Mitochondria and melanosomes establish physical contacts modulated by Mfn2 and involved in organelle biogenesis

Tiziana Daniele 1, Ilse Hurbain 2, Riccardo Vago 3, Giorgio Casari 3, Graça Raposo 2, Carlo Tacchetti 4, Maria Vittoria Schiaffino 5

Affiliations expand

- PMID: 24485836 DOI: 10.1016/j.cub.2014.01.007

Free article

Abstract

Background: To efficiently supply ATP to sites of high-energy demand and finely regulate calcium signaling, mitochondria adapt their metabolism, shape, and distribution within the cells, including relative positioning with respect to other organelles. However, physical contacts between mitochondria and the secretory/endocytic pathway have been demonstrated so far only with the ER, through structural and functional interorganellar connections.

Results: Here we show by electron tomography that mitochondria physically contact melanosomes, specialized lysosome-related organelles of pigment cells, through fibrillar bridges resembling the protein tethers linking mitochondria and the ER. Mitofusin (Mfn) 2, which bridges ER to mitochondria, specifically localizes also to melanosome-mitochondrion contacts, and its knockdown significantly reduces the interorganellar connections. Contacts are associated to the melanogenesis process, as indicated by the fact that they are reduced in a model of aberrant melanogenesis whereas they are enhanced both where melanosome biogenesis takes place in the perinuclear area and when it is actively stimulated by OA1, a G protein-coupled receptor implicated in ocular albinism and organellogenesis. Consistently, Mfn2 knockdown prevents melanogenesis activation by OA1, and the pharmacological inhibition of mitochondrial ATP synthesis severely reduces contact formation and impairs melanosome biogenesis, by affecting in particular the developing organelles showing the highest frequency of contacts.

Conclusions: Altogether, our findings reveal the presence of an unprecedented physical and functional connection between mitochondria and the secretory/endocytic pathway that goes beyond the ER-mitochondria linkage and is spatially and timely associated to secretory organelle biogenesis.

OK okay ok-kay....

We got Microtubules connected to Mitochondria, Microtubules connected to Melanosomes and now we have Mitochondria connected to Melanosomes. What further demonstration needs to happen here?

Lets do a Recap and add in some more details...

Electricity is produced by ions. This is the basis of chemistry, chemicals.

Atoms are the basic particles of the chemical elements. An atom consists of a nucleus of protons and generally neutrons, surrounded by an electromagnetically bound swarm of electrons.

A chemical compound is a chemical substance that is composed of a particular set of atoms or ions. Two or more elements combined into one substance through a chemical reaction form a chemical compound. All compounds are substances, but not all substances are compounds. A chemical compound can be either atoms bonded together in molecules or crystals in which atoms, molecules or ions form a crystalline lattice. - wiki

Nerves, Muscle Cells, Neurons etc... use chemicals to create electricity. Even the Heart Nodes use the same process of Sodium and Potassium we discussed to produce electricity.

This may be the Nexus regarding electricity and Melanin, Sodium? We still need more information on sodium and melanin though... Back to the recap.

In the Heart Nodes Calcium is coupled with Sodium to increase the "Positivity". The Action Potential strikes like lightning from the **North Western** corner of the Heart to the bottom.

Electrolytes are really Electric-Lights right? Once that lightning flashes, the Nodes are reset, just like the Neurons.... The heart is made out of Neurons, Melanocytes and Cardiomyocytes for a reason. The beat or pulse of the heart is the basic 1 and 0 of the entire body. We have covered this ad naseum in the other books.... I know, but we have to keep this in mind.

Heart Rate Variability (HRV) is the physiological phenomenon of variation in the time interval between heartbeats. It is measured by the variation in the beat-to-beat interval. -wiki

This means that the Heart and most likely the entire body run on Alternating Current.

This is interesting because the standard for AC is 60 Hertz. This is why AC is so dangerous it is based on your body... Of course we can't use 60 hertz or stick our hands in electric sockets but...

You now have a deeper explanation for spontaneous fibrillation. Fibrillation is literally the Heart losing its AlgaRhythm...

Are you studying your Time Table?

Take a moment to open the AlgaRhythm & L'Goat books...

Review questions and answers 3, 7, 13 & 16 see if anything clicks....

The idea is the Valves (zinc from our websites repairs these valves hint hint wink wink), have to open and close at different times. The current must alternate to facilitate that job amongst others...

Think about the brain you have been told your entire life that when your sleep, electricity is flowing between 3-7 hertz and being awake the juice is flowing between 8-30 hertz with blast in the gamma wave from 30 to over 100...

That is clearly ALTERNATING CURRENTS.

Gamma Waves 30 - 100 HZ	/\/\/\/\/\/\/\/\/\/\/\/\/\/\/\/\	INSIGHT PEAK EXPERIENCES SYNCHRONIZATION
Beta Waves 12 - 30 HZ	/\/\/\/\/\/\/\/\	ALERTNESS CONCENTRATION THINKING
Alpha Waves 8 - 12 HZ	/\/\/\/\/\	MEDITATION CREATIVITY RELAXATION
Theta Waves 4 - 8 HZ	/\/\/\	VISUALIZATION TRANCE DREAMING
Delta Waves 0.5 - 4 HZ	/\/\	DEEP SLEEP TRANSCENDENCE RESTORATION

We bout to bang a hard left real quick too...

Nikola Tesla was born in 1856 in Austria-Hungary and emigrated to the U.S. in 1884 as a physicist. He pioneered the generation, transmission, and use of alternating current (AC) electricity, which can be transmitted over much greater distances than direct current.

Tesla patented a device to induce electrical current in a piece of iron (a rotor) spinning between two electrified coils of wire. This rotating magnetic field device generates AC current when it is made to rotate by using some form mechanical energy, like steam or hydropower. When the generated current reaches its user and is fed into another rotating magnetic field device, this second device becomes an AC induction motor that produces mechanical energy. Induction motors run household appliances like clothes washers and dryers. Development of these devices led to widespread industrial and manufacturing uses for electricity.

The induction motor was only part of Tesla's overall conception. In a series of history-making patents, he demonstrated a polyphase alternating-current system, consisting of a generator, transformers, transmission layout, and motor and lights. From the power source to the power user, it provided the basic elements for electrical production and utilization. Our AC power system remains essentially unchanged today.

In 1888, George Westinghouse, head of the Westinghouse Electric Company, bought the patent rights to Tesla's system of dynamos, transformers and motors. Westinghouse used Tesla's alternating current system to light the World's Columbian Exposition of 1893 in Chicago. Then in 1896, Tesla's system was used at Niagara Falls in the world's first large hydroelectric plant.

The Tesla coil, invented in 1891, is still used in radio and television sets, car starters, and a wide variety of electronic equipment.

Tesla's work with radio-frequency waves laid the foundation for today's radio. He experimented with wireless transmission of electrical power, and received 112 patents for devices ranging from speedometers to extremely efficient electrical generators to a bladeless turbine still in use today. He suggested that it was possible to use radio waves to detect ships (later developed as RADAR), and his work with special gas-filled lamps set the stage for the creation of fluorescent lighting.

Tesla was Thomas Edison's rival at the end of the 19th century - in fact, he was more famous than Edison throughout the 1890's. His invention of polyphase AC electric power earned him worldwide fame but not fortune. At his zenith his circle of friends included poets and scientists, industrialists and financiers. Yet Tesla died alone and almost penniless in a New York hotel room in 1943. During his life, Tesla created a legacy of genuine invention that still fascinates today. After his death, the world honored him by naming the unit of magnetic flux density the "tesla."

Patent Number 390,414

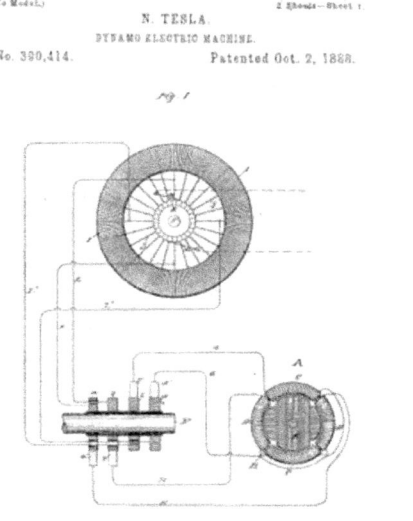

Can you now see where Tesla was getting his designs? You should immensely empowered on many levels after this book!

Are you ready to take the challenge serious now?

Using the Fiscal Edition, the Gold Book, Divine Mathematics and the AlgaRhythm to change your lifestyle.

Are you ready to take the Oath as a Human Electrician having completed the world's first Human Electrician Course?

THE WAS SCEPTRE

I believe based on the Data thus presented, our ancestors were communicating something else to us also about the Aorta, via the Was as a staff.

Our center of gravity or Center of Mass, is central to how we move, run and walk.

In Jujitsu we learn this is where strength comes from, in basketball, football & baseball I was told speed came from my core. I assumed core meant abs...

I now know that the Diaphragm Network of muscles we call our abs is situated around the Aorta, the Aorta branches right around the Hips/Pelvis region...

The Horse Stance is the Natural Shape of the Aorta...

I can't count all the hours I spent in Horse Stance focused on breathing LMMFAO... I was focused on everything but breathing...

When a human takes a hit, that is applied force. Floyd Mayweather is a master of managing applied force. Floyd never allows you to target his Center of Mass, the should roll is more than a blocking movement... A blow to your center moves you backwards and you must absorb all the force. A blow off center causes your torso to rotate to alleviate the the impact... Even when he is hit, he isn't really hit...

This gives new meaning to light on your feet...

There is a lot of information to process and expound on...

If the current in our body alternates does it make it AC?

Not exactly that is just a step in the right direction to learn. AC is typically fixed. Your body is a amalgamation of miniature super computers. The currents alternate because they are constantly encoded with new data!

Direct Current and Alternating Current are the most well known, but whats most akin to what we have is Transient Current. This is too and approximation to better understand your Electronic Super Suit...

Light creates Impulse Transients in Electrical Systems that run in 1 direction. If those 'transients' or disturbances go in multiple directions they are called Oscillatory Transients.

This maybe the process of how the Soul works...

We know about the Photoelectric Effect right?

PhotoElectric Effect - When light shines on a metal, electrons can be ejected from the surface of the metal in a phenomenon known as the photoelectric effect.

We know the basics of 1s and 0s or broken vs solid lines...

People have asked, "Enqi we believe your onto something with your discovery that the body runs on light but..."

SMDH...

How many once in a lifetime discoveries do you think 1 guy is supposed to make LMAO...

This is that interface none the less.

The Light powers and informs the body.

This should also make you think **VERY DIFFERENTLY ABOUT SYNTHETIC LIGHT**. You may be worried about the wrong things...

We have a wide variety of light sensors in the body, especially in places that are not exposed to external light. This means only one thing, they are there to absorb internally produced light.

This is what the basis of my 2018-2019 paper was about that Jack Kruse attempted to Plagiarize in 2023.

HUMAN CHEMILUMINESCENCE OF PHOSPHORUS IS THE SOURCE OF HUMAN BIOLUMINESCENCE AND HUMAN PHOTOLUMINESCENCE

Chemiluminescence

LivingBioChemistry.com
W.M. Dr. Enqi Real C.P.T., S.N., N.D., M.H.
Living Biochemistry
March 2, 2019
The entire aspect of energy production and information transfer cell to cell, molecule to molecule, enzyme to enzyme and across has overlooked Phosphorus in its central role to Human Health hence has been misinterpreted and/or misrepresented as Sugar, Protein or Electron driven via Water. Phosphorus in Phosphorylation (adding phosphorus) and De-Phosphorylation, the Cori cycle and the Tricarboxylic Acid Cycle (Kreb cycle); Phosphorus in DNA/RNA cycles, that is Methylation; the Phospholipid cycles (metabolism and synthesis of fats/All cell membrane water/minerals interactions) are all Chemiluminescent via the Attraction & Repulsion of photons to and from Phosphorus.

The biochemistry that is life is powered by ATP (sugar/amino/3 phosphoruses) and the firing off of a Phosphorus in the Cytoplasm releases light that is used to power the cell. Fat is known as our STORED ENERGY because of its sugar content however the light producing PHOSPHORUS (phosphates) attached to the sugar molecules is overlooked. Information is transferred throughout the body via the many light-producing mechanisms elucidated here and many more we will not disclose yet that is the system of crystals and pigments in the cytosol in conjunction with the ascia and interstitium. Every cell uses its nDNA as an antenna, and the body uses its spine in the same manner. Sound waves encode information and transformed via proteins into "light" which is sent out immediately or in the molecule which is one of the main mineral functions, frequency focus much like the crystals in radio tuned into a radio station. The entire science of the Human Body must be re-written with Water, Melanin and Light as it's core.

Reference Materials

Eat Right 4 Your Haplotype written by Dr. EnQi ReaL

IKing Squared written by Dr. EnQi ReaL

IKing Cubed the Creation of Man hence written by Dr. EnQi Real

We have come very far since then however the rest of the world has yet to even catch up....

Plasma based Crystal disc, fitted with integrated circuits as well as gates and channels fitted in a Capacitor (membrane). The full body capacitors are also fitted with a wide variety of antenna and transformers. These Nanobots have 2000+ alternators and motors in them for internal power. Each of these cells are Electrochemical with thousands of smaller nano electric machines.

We now have to add the filaments into this Crystal Disc we call a Body Cell or Somatic Cell. It make sense and it's easy to understand what they are doing in Neurons, the amplify the electromagnetic signal generated in the cell body, until it can be translated into chemicals. Got it! In the dendrite they perform the task just backwards, they allow the incoming chemicals to be translated back into electromagnetic signals for the cell body. Digital/Analogue Binary Coding...

We are now adding the light from GTP & ATP in to the equation. Oxidize phosphates & Reactive Species being released, form a steady stream of information. Melanins, Globulins, Opsins etc...

The pigments are usually in cells that have nerves attached but not hard wired themselves. In the cells we now know they are working with the microtubule system.

ATP & GTP

Isn't it odd these guys are from the DNA?

Adenosine Tri-Phosphate - Adenine (binds to Thymine in RNA it binds to Uracil)

Guanosine Tri-Phosphate - Guanine (binds to cytosine)

The A in FAD & NAD is also this guy too!

The FAD that works via Cytochrome 2 and the NAD from the EnQi Cycle & Cytochrome 1...

This maybe the nexus Nucleosides and Nucleotides in our Nutrition...

I wonder if.... hmmm....

Both adenine and guanine are derived from the **nucleotide** inosine monophosphate (IMP), which in turn is synthesized from a pre-existing **ribose phosphate** through a complex pathway using atoms from the amino acids **glycine, glutamine, and aspartic acid**, as well as the coenzyme **tetrahydrofolate**.

This is ribose, phosphorus, oxygen, 3 amines and 'folate'??? Obviously not that simple but this is how you investigate in EnQi ViSion!

This will lead us over to seeing more info on Ribose Phosphate right?

The pentose phosphate pathway (also called the phosphogluconate pathway and the hexose monophosphate shunt and the HMP Shunt) is a metabolic pathway parallel to glycolysis.[1] It generates NADPH and pentoses (5-carbon sugars) as well as **ribose 5-phosphate**, a precursor for the synthesis of nucleotides.[2] While the pentose phosphate pathway does involve oxidation of glucose, **its primary role is anabolic** rather than catabolic. The pathway is especially important in red blood cells (erythrocytes). The reactions of the pathway were elucidated in the early 1950s by Bernard Horecker and co-workers.[3][4]

There are two distinct phases in the pathway. The first is the oxidative phase, in which NADPH is generated, and the second is the non-oxidative synthesis of 5-carbon sugars. For most organisms, the pentose phosphate pathway takes place in the cytosol; in plants, most steps take place in plastids.[5]

Like glycolysis, the pentose phosphate pathway appears to have a very ancient evolutionary origin. The reactions of this pathway are mostly enzyme-catalyzed in modern cells, however, they also occur non-enzymatically under conditions that replicate those of the Archean ocean, and are **catalyzed by metal ions**, particularly ferrous ions (Fe(II)).[6] This suggests that the origins of the pathway could date back to the **prebiotic world**. - wiki

This is why the learning never ends...

The Oath is located in L'Goat book on pages 305-310.

To begin you need on the job training. That training needs to be documented by completing the Divine Mathematics Book Charts and the AlgaRhythm Book Charts, then writing your first book as a Certified Human Engineering Graduate!

www.ingramcontent.com/pod-product-compliance
Lightning Source LLC
Chambersburg PA
CBHW052151220526
45471CB00004B/1630